Boeing and Chicago
share a great
history together
(p. 15 →)

[signature]

John Zukowsky

TO THE MEN AND WOMEN WHO HAVE RISKED THEIR LIVES—AND
SOMETIMES MADE THE ULTIMATE SACRIFICE—FOR THE ADVANCEMENT
OF FLIGHT AND THE PURSUIT OF THE UNKNOWN.

EDITED BY
ANTHONY M. SPRINGER

AEROSPACE DESIGN

Aircraft, Spacecraft, and the Art of Modern Flight

MERRELL
LONDON · NEW YORK

AEROSPACE DESIGN

CONTENTS

PREFACE

On a stretch of Florida coastal marshlands, magnificent blue herons, bald eagles, and sandhill cranes gracefully glide on ocean-warmed breezes past powerful rocket boosters poised for their fiery ascents into space.

A product of sublime design, birds like these have inspired humanity's dream of flight since time immemorial, a dream realized a century ago above the coastal dunes of North Carolina, and one that now continues from NASA's John F. Kennedy Space Center.

Since the Wright brothers' first powered flight at Kitty Hawk, a number of dreamers and doers have helped to pioneer the air and space frontier, with the men and women of NASA proudly playing a prominent role.

Fittingly, the airplanes, boosters, and space vehicles produced by NASA and its predecessor organization, the National Advisory Committee on Aeronautics (NACA), reflect an unrivaled passion for excellence and creativity in design. This volume celebrates that passion—and the artistry ensuing from our scientists' and engineers' efforts to expand our horizons into the realm of birds and beyond.

The materials presented in this volume have been painstakingly gathered from the storehouses, displays, laboratories, and backyards of NASA's far-flung installations and from our government and industry partners.

When you gaze at the objects and artifacts pictured in this book, I encourage you to visualize the talent, ingenuity, creativity, and dedication it took to conceive, develop, manufacture, test, and fly these wonders of the modern world.

NASA's ambition for the future, as set out in our statement of bold mission goals, is to understand and protect our home planet, explore the Universe, search for life, and to inspire the next generation of explorers … as only NASA can.

I hope that this volume will play a part in inspiring the next generation of explorers and reminding everyone who has ever experienced the thrill of flight or has been filled with pride at our space exploration achievements that the sky is no longer the limit.

Sean O'Keefe
Administrator
National Aeronautics and Space Administration

1. A cavalcade of flight: this exhibit represents the advances made in flight during its first seventy-five years. Each vehicle, from the Wright Flyer of 1903 to Space Shuttle, under design in 1978, is represented by a model. Recent decades have seen improvements to existing designs, the debut of a new generation of airliners, stealth technology in fighters and bombers, and the widespread use of fly-by-wire in every type of aircraft.

FOREWORD

You know you've achieved perfection in design,
not when you have nothing more to add,
but when you have nothing more to take away.

Antoine de Saint-Exupéry, *The Little Prince* (1943)

For all of human history, man has been envying the birds' ability to fly. History does not record when toy gliders and kites were first flown. The nineteenth century, however, produced a number of inventors who developed winged flying models, kites, and gliders capable of lifting a person. It was natural to emulate the flight of the birds. Wings could be built that would allow gliding flight, but all the attempts to build flapping wings were unsuccessful and, in some cases, disastrous.

Wilbur and Orville Wright experimented with kites and gliders, and were successful in being able to determine the relationships between the size of a wing, the airfoil, the wind speed and the weight that could be carried. They could not devise a method of flapping wings but did recognize that an engine-driven propeller could be used to provide the force necessary for powered flight. Unfortunately, there were neither propellers nor the technology to design them, nor was there an appropriate engine available to purchase. So they designed and built both the engine and the propellers. They recognized that a successful flying machine would need to be controlled in all three axes—pitch, roll, and yaw—and devised a

system to provide such control. On December 17, 1903, the Wrights made four successful flights in their Flyer at Kill Devil Hill, North Carolina. They were the first successful aircraft designers.

Less than twelve years later, on March 3, 1915, the Congress of the United States, recognizing the importance of the burgeoning aircraft industry, formed the Advisory Committee for Aeronautics (later National Advisory Committee for Aeronautics, or NACA) "to supervise and direct the scientific study of the problems of flight, with a view to their practical solution." After the launch of the first man-made satellite, Sputnik, by the Soviet Union in 1957, the Congress established the National Aeronautics and Space Administration (NASA) with the NACA as its core.

Throughout the past century, NACA and NASA have been at the cutting edge of aeronautical and space development. *Aerospace Design* is the exciting and fascinating story of many of the contributions of NACA and NASA researchers to the memorable advancements in the world of flight during the twentieth century. We can only hope that the progress in the twenty-first century will be equally remarkable.

Neil Armstrong
Former NASA Research Pilot, Astronaut,
and NASA Deputy Associate Administrator
for Aeronautics

2. NASA pilot Neil Armstrong stands next to X-15 ship #1 after a research flight in 1960. A total of 199 flights would be made during the X-15 program between June 1959 and October 1968 using the three flight vehicles. These would result in the fastest flight (Mach 6.7) and highest flights of an aircraft until the entry of the Space Shuttle orbiter in 1981. Neil Armstrong flew seven X-15 flights before beginning his training as an astronaut.

John Zukowsky

INTRODUCTION

3. The Propeller Research Tunnel, built for the National Advisory Committee for Aeronautics (NACA), Langley, Virginia, by the Austin Company in 1925. The photograph shows the Sperry M-1 Messenger being tested in 1927. Architectural firms such as the Austin Company have developed a strong reputation for building aviation facilities throughout the US from the 1920s to the present day.

When most people think of aerospace design they automatically think of engineers, scientists, and technicians; they do not initially see architects and designers as having contributed much to this field. Yet, as some of the books and exhibitions that I have been involved in recently have demonstrated, visual arts professionals have had as much to do with the development and overall image of aviation and space travel in the last century as have aerospace engineers.[1] The public's perception of the aerospace engineer is most probably one of the detached scientist searching for the objective truth about flight. Recent literature, however, has shown how the work of scientists and engineers can be as subjective as the work of those involved in the fine and literary arts.[2] Yet people still tend to think of engineers as providing practical solutions to problems rather than aesthetic ones, which are supposedly in the realm of architects and designers.

This book, and its associated exhibition, aims to challenge such preconceptions and demonstrate

that aerospace design is more than a matter of nuts, bolts, and rivets. It illustrates the important role played by industrial designers as well as aeronautical specialists in creating the conceptual aircraft of the future. It shows how over the past century architectural firms have participated with aerospace engineers in the design and construction of technical facilities for aeronautical research (fig. 3). It proposes that objects made for aeronautical engineering purposes, such as wind tunnel models, as well as the aircraft themselves, are strikingly aesthetic, and can have as much visual impact as did the propellers and gears shown in the landmark *Machine Art* exhibition held at the Museum of Modern Art, New York, in 1934.

As well as illuminating these issues, the essays in this book combine to present a survey of aerospace design as reflected in the work of the National Advisory Committee for Aeronautics (NACA), which was founded in 1915, and its successor the National Aeronautics and Space Administration (NASA), founded in 1958 during

4. Fantasy Photomontage Souvenir of the Chicago Aviation Meet, dated August 1914. The same view was used for a variety of early aviation meets in Chicago, and was published at various scales, ranging from postcards to posters.

the Cold War to counter the Soviet Union's early lead in the space race following its 1957 launch of Sputnik. As a prelude to these essays, however, it will be helpful to provide an overview of the architectural and design context—both national and local—to this project, especially in relation to The Art Institute of Chicago and its role in helping to bring aeronautical as well as architectural history to the general public.

AVIATION IN THE MIDWEST—THE EARLY YEARS

When one considers the architectural and engineering monuments of the US aerospace industry one naturally thinks of the east or west coasts: the great aircraft factories in the Los Angeles or Seattle areas, or the space launch facilities in California and Florida. Yet the Midwestern region of the United States has made great contributions to the history of aeronautics. These stretch back to the experiments of aviation pioneer Octave Chanute (1832–1910), author of the important 1894 publication *Progress in Flying Machines*, and his participation in the International Congress on Aerial Navigation held in conjunction with Chicago's World's Columbian Exposition of 1893. This was one of the many congresses that were organized in conjunction with the fair, most—if not all—of which were held in The Art Institute of Chicago's 1893 building.

We should recall that the Wright brothers, although making their historic first flight in Kitty Hawk, North Carolina, were Midwesterners who were in continuous contact with aeronautical pioneers such as Chanute. Their shop in Dayton, Ohio, and related facilities have been restored and have become shrines to the brothers' ingenuity. South of the museum, in Grant Park, Chicagoans witnessed a spectacular International Aviation Meet from August 12 through 20, 1911 and over several subsequent years; the meet was the subject of many postcards and fanciful promotional items (fig. 4).

As was the case with other US cities and states, Chicago and Illinois played their part in World War I. The War Department established an Air Corps training base in Rantoul, Illinois, some 15 miles (24 km) north of the University of Illinois in Urbana-Champaign, where related ground-training was done. The Rantoul base was opened in July 1917 and later in the century functioned as Chanute Air Force Base before closing in 1993. Chicago architect John Wentworth served in the US Air Service with ace Captain "Eddie" Rickenbacker in the famous 94th Aero Squadron. Rickenbacker recounts how he and several officers concocted the "hat-in-the-ring" emblem for their squadron, with Wentworth as designer of

INTERNATIONAL AVIATION M
GRANT PARK, CHICAGO.
August 1914

it, creating one of the most famous squadron symbols in American history (fig. 5).[3] After the War, Wentworth went on to become a partner in the successful firm of Rebori & Wentworth, known for its Art Deco buildings in Chicago.[4]

The post-World War I era witnessed a number of private airports being built in the area, such as the Curtiss-Reynolds Airport in Glenview, Illinois, of 1929, which hosted the National Air Races in 1930. The city's first municipal airport (which was to become Midway Airport) was built in 1927. Such magnates as Colonel Robert R. McCormick—publisher of the *Chicago Tribune* and builder of its famous Gothic skyscraper (1922–25)—adopted the private airplane as a favored means of transport at this time. McCormick commuted in one of his Sikorsky amphibians (fig. 6) from his estate in Wheaton, Illinois, to his office some 30 miles (48 km) east near Chicago's lakefront, as well as to his estate in South Carolina and his paper-pulp mills in

Canada. (Later he was to buy a deactivated World War II B-17 bomber for his personal transport.) McCormick was well known for the aviation publicity stunts that he used to promote his newspaper, and for his championing of a lakefront airport, an idea that was considered along with the preliminary designs for a 1933 World's Fair in Chicago but not realized until Meigs Field opened in 1948.[5]

Although the 1937 and 1939 World's Fairs in Paris and New York, respectively, may have boasted the first large freestanding buildings devoted solely to aviation, it was at the 1933–34 Century of Progress Exposition on Chicago's lakefront that aviation was first comprehensively featured in a fair's overall program. (The lakefront site, not far from that of the famed 1911 Aviation Meet, was later to become the location of the now-closed Meigs Field.) The program included fly-bys within a "Wings of a Century" pageant, which surveyed the rapid advances in transportation over the

5. Eddie Rickenbacker in his Spad XIII with the 94th Aero Squadron emblem, 1918. Rickenbacker, America's leading ace in World War I, became the head of Eastern Airlines in 1938 and was their chairman from 1954 to 1963.

7. A Boeing 247 on display in the Travel and Transport Building, Century of Progress Exposition, Chicago, 1933. The epitome of streamlined air travel at the fair, the Boeing 247 was seen alongside other examples of streamlining in transportation at the time, from Buckminster Fuller's Dymaxion Car to the Burlington Zephyr railroad train.

6. Col. Robert R. McCormick (right) exiting his Sikorsky amphibian, the Untin Bowler. "The Colonel," as he is often called, was an ardent aviation enthusiast and an early aero-commuter.

previous hundred years. (A historical exhibit was planned, though apparently not executed, in a small building called the "Airshow" just north of the Travel and Transport Pavilion.) More important, the fair's Travel and Transport Pavilion afforded visitors a close look at one of United Airlines' new streamlined Boeing 247 airliners—and for those who could not attend the fair Boeing toured a 247 transport throughout the United States (fig. 7). Technically minded visitors would have appreciated the streamlined engine cowlings and variable pitch propellers that owed their existence to NACA research; and all would have admired the plane's overall streamlined metal design and construction, as well as its capabilities to move passengers in air-conditioned comfort at speeds in excess of "three miles a minute."

The culmination of aviation activities at the fair came when, on July 15, 1933, more than a million spectators witnessed the arrival of a squadron of twenty-four Savoia-Marchetti S-55X seaplanes commanded by General Italo Balbo. They landed on the lake after a 6100 mile (9818 km) journey from Italy. The arrival of the planes coincided with the unveiling of a gift from Mussolini to Chicago—a column from a Roman temple in Ostia, which sits on the lakefront opposite Soldier Field even today.[6]

WORLD WAR II AND THE GROWTH OF COMMERCIAL FLIGHT

The USA's entry into World War II brought dramatic aviation changes into Chicago. These changes range from temporary ones, such as increased industrial facilities to produce goods in support of the war effort, to a long-term permanent impact on the facilities and demographics of a region. On an ephemeral though important historic note, between 1943 and 1945 Grumman Wildcat fighters, along with Douglas Dauntless and Grumman Avenger bombers, were used by naval aviators to practice take-offs and landings on training aircraft carriers in Lake Michigan (fig. 8). Two such carriers were the *Wolverine* and *Sable*, converted paddle steamers located off the shores of Chicago, not far from Glenview Naval Air Station (originally Curtiss-Reynolds Airport) where the naval aviators were based. (George H. Bush, later to become US President, was one of the more famous aviators to have trained here.) Nowadays the lake bottom is littered with some two hundred wrecks, a number of them from planes that were deemed expendable after they had served their purpose, others the result of trainee pilots missing their approach.

Airfields and related structures of a more permanent nature were built on land near to Chicago and throughout the Midwest. The Curtiss-Reynolds Airport of 1929 in Glenview, Illinois, with its now landmark hangar, became the Glenview Naval Air Station in 1940, serving as such until its closure in 1995 following the end of the Cold War. Even more important were the war efforts of such architectural firms as Michigan-based Albert Kahn Associates and the Austin Company of Cleveland, which designed spectacular buildings within President Franklin Delano Roosevelt's so-called "Arsenal of Democracy."[7] These included aircraft- and aviation-related factories throughout the United States, including the Chicago region, all of which contributed to the US's massive wartime production of more than 300,000 airplanes. Some of the most famous Kahn buildings in that regard are the still-extant 1942 Ford Plant in Willow Run, Michigan (now part of General Motors), which produced B-24 Liberator bombers, and the 1942 Dodge Chicago factory, which made engines for B-29 Superfortress bombers. Additional wartime aviation factory complexes in the Midwest include

8. Grumman Avenger landing on USS *Wolverine*, with Landing Signal Officer Lt. C.E. Roemer on the deck, April 10, 1943. Pilots based at nearby Glenview Naval Air Station used the *Wolverine*, a converted lake paddle-wheeler, for carrier qualification training on Lake Michigan, far away from Axis submarines off the Atlantic and Pacific coasts.

the Republic plant in Evansville, Indiana, which assembled P-47 Thunderbolt fighters, and the Curtiss factory in Columbus, Ohio, which produced Helldiver torpedo bombers, as well as the Boeing factory in Wichita, Kansas, which assembled B-29s.

More important for our purposes, however, was a Chicago aircraft factory that was significant not only for its war use but also for its post-war development. The Austin Company, known for its many industrial buildings, including those for numerous Boeing plants around the nation, constructed the Douglas Factory at Orchard Field, northwest of Chicago, in 1943. When built it was said to be the world's largest wooden factory, and it extensively utilized other non-strategic materials as well. The factory built C-54 transports, which, at the end of the War, were often recycled into commercial airliners and air freighters—the latter becoming one of the symbols of the Berlin Airlift of 1948, the so-called "Rosinenbombers" or "Raisin Bombers" (fig. 9).

Following the end of World War II, the city of Chicago purchased the airfield and adjacent land from the federal government in 1946 and, in 1949, named the new commercial airport in memory of Illinois resident and World War II ace Lt. Commander Edward "Butch" O'Hare, who shot down five Japanese bombers on February 20, 1942. In 1949, the city council also renamed the newly finished terminal and remodeled municipal airport "Midway" to commemorate the important aerial naval battle of June 4–7, 1942, which turned the tide of the war in the Pacific. In its busiest year of 1959, Midway handled ten million passengers. (O'Hare, which officially opened to commercial traffic in 1955, only had about two million passengers in 1959.) After the opening in 1963 of O'Hare's new buildings—which were purposely constructed to accommodate the new jet airliners of the era—the airport's passenger numbers soared until it became the world's busiest airport (a title it maintained for many years, before ceding to Atlanta's Hartsfield Airport). Despite this extensive use, and subsequent additions and remodelings, the post-war plan and jet-age terminals of O'Hare still function well: the airport carried 66 million people during 2001 in more than 911,000 take-offs and landings.

As well as its important airports, Chicago has another claim to having made commercial aviation history. In 1949 the city was host to the Convention on International Civil Aviation, which established the International Civil Aviation Organization (ICAO). This body has been responsible for worldwide commercial air transport policy ever since.

It is clear, then, that over the past century Chicago and the surrounding region have made significant contributions to the story of aviation in the USA—even before Boeing moved its corporate headquarters here in 2001. But what role has The Art Institute of Chicago played in disseminating this story?

ART INSTITUTE EXHIBITIONS
Of Chicago's museums, it is the Museum of Science and Industry that people would perhaps most readily link with aeronautics. Indeed, in 1996 the museum opened an important aviation display, complete with a United Airlines Boeing 727 as its centerpiece—a display that made news headlines with the airliner being towed off Lake Shore Drive into the building. The museum's permanent collection has examples of some rare aircraft, including a Junkers Ju-87 Stuka, one of the very few survivors of this type. However, since the late 1980s the Department of Architecture at The Art Institute of Chicago has also been an active participant in aviation and architectural history; it was at this time that grants from the National Endowment for the Arts helped the department to expand its definition of collecting and exhibiting architecture by incorporating aspects of the designed environment created by urban planners, landscape architects, industrial designers, and graphic designers.

With major support from the National Endowment for the Humanities, the Department of Architecture created two powerful products from this new holistic approach to design. The first of these was the 1996 exhibition *Building for Air Travel: Architecture and Design for Commercial Aviation*. Helmut Jahn—architect for the striking United Airlines terminal at O'Hare (1983–88) as well as recent airports and airport buildings in Cologne and Munich in Germany, and Bangkok, Thailand— designed the installation of aluminum ribs, creating an environment that resembled a stylized airplane under construction (fig. 10). Jahn's simple

9. C-54 transport under assembly in the Douglas Factory at Orchard Field, north-west of Chicago, 1943. The Douglas Factory served as the basis for today's O'Hare International Airport. The world's busiest airport in terms of operations, it handled more than 922,000 flights in 2002.

yet powerful design reminded visitors that all the aviation work that has been done by architects and designers would not exist at all were it not for the invention of the airplane itself. The exhibit not only surveyed airports, aircraft factories, and related maintenance facilities but also showed the development of aircraft interiors and airline corporate identity, stretching back to the creation of Lufthansa's crane logo in 1919 by Berlin architect and German wartime aviator Otto Firle. The exhibition traveled to other museums as well as to airports around the world, and was the Department of Architecture's second most popular exhibition.

Edging this one out in attendance numbers was another exhibition devoted to an aerospace theme, entitled *2001: Building for Space Travel*. This 2001 exhibition, co-organized with the Museum of Flight in Seattle, focused on the work that architects, designers, and even film production designers have had in shaping our image of space flight, in reality as well as in science fiction. It included the architecture of actual space-launch facilities as well as the work done by set designers on fantasy spacecraft, such as those designed by production designer Herman Zimmerman for recent *Star Trek* films and television shows. It also included work done by such industrial designers as John Frassanito (born 1941). He is a Houston-based designer who first worked for Raymond Loewy on Skylab, America's first space station (1973), and more recently participated in the conceptual designs for the current International Space Station.[8]

Since the early 1990s Frassanito has become one of the leading visualizers—through his computer animations—of future spacecraft and aircraft for NASA (see Chapter 7, figs. 132, 133, 142–44, 148). For the *Building for Space Travel* exhibition Chicago architect Douglas Garofalo created an anthropomorphic, organic-looking environment. Garofalo is an aficionado of the expressive possibilities of today's computer-design programs, and his design incorporated pedestals that evoked the terrain of another planet, alongside angular steel ribs with irregularly pulled metallic cloth on the opposite wall, which created an otherworldly environment and projected an impression of rather threatening, oversized, science-fiction-style creatures (fig. 11). As in the case of *Building for Air Travel* the exhibition traveled to other

10. Installation of *Building for Air Travel* at The Art Institute of Chicago, 1996. Noted architect Helmut Jahn designed this exhibition of stylized aluminum ribs to recall the image of an aircraft under construction.

museums, and a variant exhibition was shown in airports and other public places.

Both of these interrelated aerospace exhibitions laid the ground for our current exhibition on aerospace design. As with those two earlier examples, this exhibition features a creatively contextual installation—in this instance prepared by Chicago architect Jeanne Gang of Studio Gang—along with a variant shown at public places such as airports.

THE PERMANENT COLLECTION

As well as staging temporary exhibitions, the Department of Architecture has also expanded

11. Installation of *2001: Building for Space Travel* at The Art Institute of Chicago, 2001. Through computer-aided design, Chicago architect Douglas Garofalo created a variety of distinctive shapes and structures that were intended to recall an environment beyond the Earth.

12. Vornado fans, past and present. Designed by Richard Ten Eyck, the Vornado borrows from aerospace technology, particularly the way in which the fan moves air in a concentrated cone much like a jet engine. Ten Eyck worked extensively on the design of Cessna airplanes and Bell helicopters for much of his career in the 1960s and 1970s.

the limit of its collections to include not only architects' drawings for airport buildings but also the work of industrial designers for aerospace. One of the most important, yet least recognized, of these designers is Richard Ten Eyck (born 1920). Raised in Marseilles, Illinois, and trained in Chicago, Ten Eyck moved to Wichita in 1945 to work with the Beech Aircraft Company on the Model 35 Bonanza (1947). He established his own design firm in 1948 and went on to act as design consultant for a number of large American companies, including Westinghouse and Hesston Tractors, and also spent much time as a stylist and designer for Cessna Aircraft (1950–80) and Bell Helicopters (1980–90). However, it is for the Vornado fan that Ten Eyck is best remembered (designed for the O.S. Sutton Corporation c. 1945–59; variants subsequently designed for Vornado Air Circulators, Inc.) (fig. 12). Ten Eyck's creation was consciously inspired by the new jet-engined airplanes of the mid-1940s: unlike

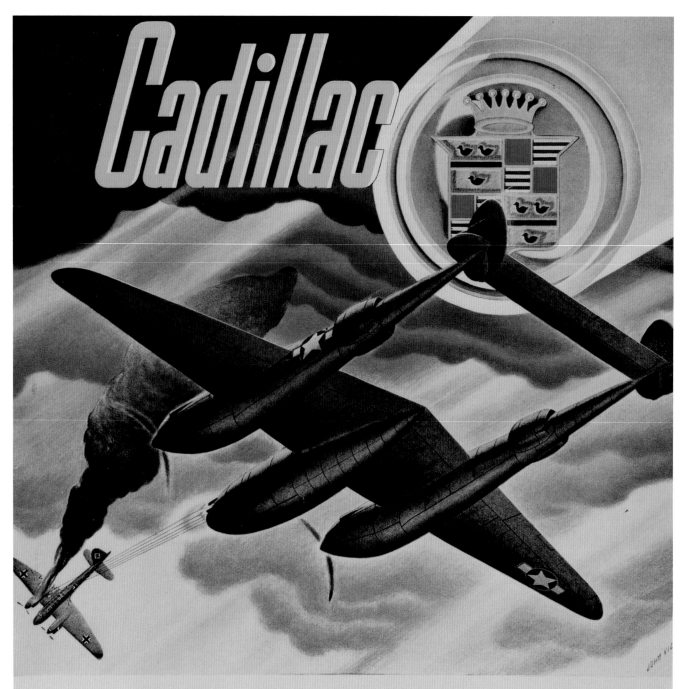

Craftsmanship is still our stock in trade

The rhythmic roar of the P-38 tells more eloquently than words of the superb fighting qualities built into its two perfectly synchronized engines. Foremost of the American designed and built liquid-cooled aircraft engines is the Allison, which powers several of our top fighter craft and for which we at Cadillac produce vital precision assemblies.

It was natural that Cadillac should be entrusted with this war production assignment,

because for forty years Cadillac has exemplified the ultimate in craftsmanship and precision. The long-remembered Cadillac motto, "Craftsmanship a Creed—Accuracy a Law," is far from being an empty, meaningless phrase. It is, in fact, the very credo by which we live because it calls for the fullest exercise of our highest traditional skill.

Another assignment is the production of M-5 light tanks, for which the Cadillac automotive-

type V-8 engines were adapted. This serves to keep the same Cadillac craftsmen on the same production line on which they worked in time of peace.

Thus, while serving the nation at war on a full-time basis, we are also maintaining at an efficient peak everything that the Cadillac name and crest represent in time of peace— the peace which must ultimately be ours.

CADILLAC MOTOR CAR DIVISION  GENERAL MOTORS CORPORATION

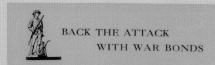

BACK THE ATTACK
WITH WAR BONDS

13. A P-38 Lightning airplane features in this Cadillac magazine advertisement from the World War II era. The twin-tailed P-38 was the brainchild of aerospace engineer Clarence "Kelly" Johnson, who was also behind the creation of a number of Lockheed's famous aircraft, from the Constellation airliner and the P-80 Shooting Star (America's first jet fighter at the end of World War II) to the famous Cold War spy planes such as the U-2 and SR-71 Blackbird.

White Sidewall Tires available at additional cost.

PICTURED here is a representative model from the finest line of Cadillacs ever built. Never in Cadillac's long and eventful history have its cars been so beautiful, so luxurious, or so magnificently engineered as they are today. It has been several years since Cadillac produced a wholly new line of cars. During that period, however, Cadillac never lost sight of its position as standard of the world for motor car quality. In keeping with that tradition of leadership, Cadillac now presents improvements which would normally be apportioned over more than half a decade. Naturally, these cars are a revelation—even to those who know Cadillac best. We believe you should see them and inspect them regardless of whether you expect to make your next motor car a Cadillac. In design, in performance and in engineering they are the world's new yardstick for quality—and as such should be of interest to everyone. Your Cadillac dealer cordially invites you to visit his showroom and inspect these magnificent motor cars.

★ CADILLAC MOTOR CAR DIVISION ★ GENERAL MOTORS CORPORATION ★

14. Cadillac magazine advertisement from 1948. Harley Earl, General Motors' chief designer, used the P-38's twin tails as inspiration for the Cadillac's tails. Soon afterward he went on to head the team that created the famous Corvette sports car for Chevrolet.

traditional electric fans, the Vornado pushes air in a concentrated cone instead of dispersing it. Much like Harley Earl's oft-cited adoption of a P-38 fighter's twin tailfins for his 1948 Cadillac (figs. 13, 14), Ten Eyck's Vornado demonstrates a clear relationship to aeronautical design of the era.

Connections such as this between architecture, design, and the aerospace industry have also helped us to reinterpret precisely an item in The Art Institute's permanent collection, as well as a related item in another museum. This concerns the famous 1942 conceptual project for a concert hall by Ludwig Mies van der Rohe. The original photomontage is in the Museum of Modern Art, New York, while The Art Institute of Chicago has a 1946 variant by Mies's student Daniel Brenner (fig. 15). Both versions use as their starting point an enlarged photograph of the 1937 Martin Aircraft Factory near Baltimore by Albert Kahn (fig. 16). Evidently, Mies and his students were impressed by the clear 300 foot (91 meter) span of this factory. Manufacturer Glenn Martin built the plant in order to make seaplanes that would be even larger than his famous Martin M-130 "China Clipper," a variant of which is shown in the original 1937 view of the plant.

AVIATION AND AESTHETICS

As the above discussion demonstrates, an understanding of Chicago's role in aviation history can greatly contribute to our understanding of the region's architecture and design. *Aerospace Design* intends to record this history, but it also goes further. We propose that engineering decisions and forms have aesthetic consequences and interrelationships with the more "artistic" forms of design, and that they are not necessarily purely scientific, purely functional, or purely objective. Likewise, we point out that visual arts professionals such as architects and industrial designers are actively engaged in working with their scientific and engineering-oriented colleagues, and that they often draw engineered forms into their own design aesthetic.

We hope that this volume will encourage the reader to discover many more interconnections between the worlds of technology and design, science and art, and learn about NASA's integral role within this symbiotic process throughout the past century of controlled flight.

15. Daniel Brenner's photomontage of a concert hall interior, 1946. This famous photomontage was based on a similar montage created in 1942 by Mies van der Rohe, Brenner's teacher. The large-span space impressed Mies and his students, and also prefigured the subsequent remodeling of many wartime factories into post-war malls and shopping centers.

16. Interior of the Martin Aircraft Factory, 1937. This view served as the basis for the well-known montages made by Mies van der Rohe and his students in the 1940s (see fig. 15). In the center of the factory, which was designed by America's most famous industrial architect, Albert Kahn, is a variant, sold to the Soviet Union, of the Martin M-130 "China Clipper." Surrounding it are a number of smaller Martin bombers, part of the company's successful export trade in the later 1930s.

Notes

1. John Zukowsky, ed., *Building for Air Travel: Architecture and Design for Commercial Aviation*, Munich (Prestel Verlag) 1996; and *2001: Building for Space Travel*, New York (Harry N. Abrams) 2001.

2. See, for example, Eugene S. Ferguson, *Engineering and the Mind's Eye*, Cambridge, Mass. (MIT Press) 1993.

3. Captain Edward V. Rickenbacker, *Fighting the Flying Circus*, Chicago (The Lakeside Press) 1919, new edn. W. David Lewis, Chicago (R.R. Donnelley & Sons) 1997, p. 36; and Charles Woolley, *The Hat in the Ring Gang: The Combat History of the 94th Aero Squadron in World War I*, Alglen, Pa. (Schiffer Publishing Ltd.) 2001, p. 28.

4. For the works of Andrew Rebori, many done while in partnership with John Wentworth, see the *Chicago Architectural Journal*, vol. 4 (1984), pp. 8–24. Wentworth's obituary is in the *Chicago Tribune* (June 21, 1958), sect. 3, p. 12.

5. According to an article in the *Chicago Tribune* (June 7, 1949), McCormick was the first aerial commuter consistently to use Meigs Field, flying his Cessna 195 from the new airport to his home in Wheaton on a fourteen-minute flight. Meigs Field was suddenly closed on March 21, 2003, by Chicago's mayor, Richard M. Daley.

6. For aviation architecture in Chicago, see John Zukowsky, ed., *Chicago Architecture and Design: 1923–1993*, Munich (Prestel Verlag) 1993, especially the essays by Brodherson (pp. 75–97) and Doordan (pp. 219–31). For aviation at the 1933 fair, see *Chicago: A Century of Progress*, Chicago (Marquette Publishing Co.) 1933, pp. 66–67 and 127; also "Huge Crowds Greet New Transport on Nation-Wide Tour," and "247 Takes Place at World's Fair," *Boeing News*, vol. IV, no. 5 (May 1933), pp. 1–4. The only summary of aviation in Chicago and its environs is Howard L. Scamehorn, *Balloons to Jets: A Century of Aeronautics in Illinois, 1855–1955*, Chicago (Henry Regnery Co.) 1957. On McCormick and his fascination with aviation see Richard Norton Smith, *The Colonel: The Life and Legend of Robert R. McCormick, 1880–1955*, Boston and New York (Houghton Mifflin Co.) 1997, pp. 268–69; P.J. Capelotti, *Explorer's Air Yacht: The Sikorsky S-38 Flying Boat*, Missoula, Mont. (Pictorial Histories Publishing Co.) 1995, pp. 4, 7, 43–48, 52, 61, and 63; and "Chicago Tribune Planes Make Flying History: Famous B-17 Has Been Sold After World Travels that Took Her to 80 Countries," *Chicago Sunday Tribune*, July 27, 1952.

7. The term "Arsenal of Democracy" comes from a speech made by President Franklin Delano Roosevelt on December 29, 1940, in which, as a response to the signing of the Tripartite Pact among the Axis powers of Germany, Italy, and Japan on September 27, 1940, he declared American support for British resistance to those dictatorships: "We must be the great arsenal of democracy. For us this is an emergency as serious as war itself. We must apply ourselves to our task with the same resolution, the same sense of urgency, the same spirit of patriotism and sacrifice as we would show were we at war. We have furnished the British great material support and we will furnish far more in the future."

8. John Zukowsky, *Space Architecture: The Work of John Frassanito & Associates for NASA*, Stuttgart (Edition Menges) 1999.

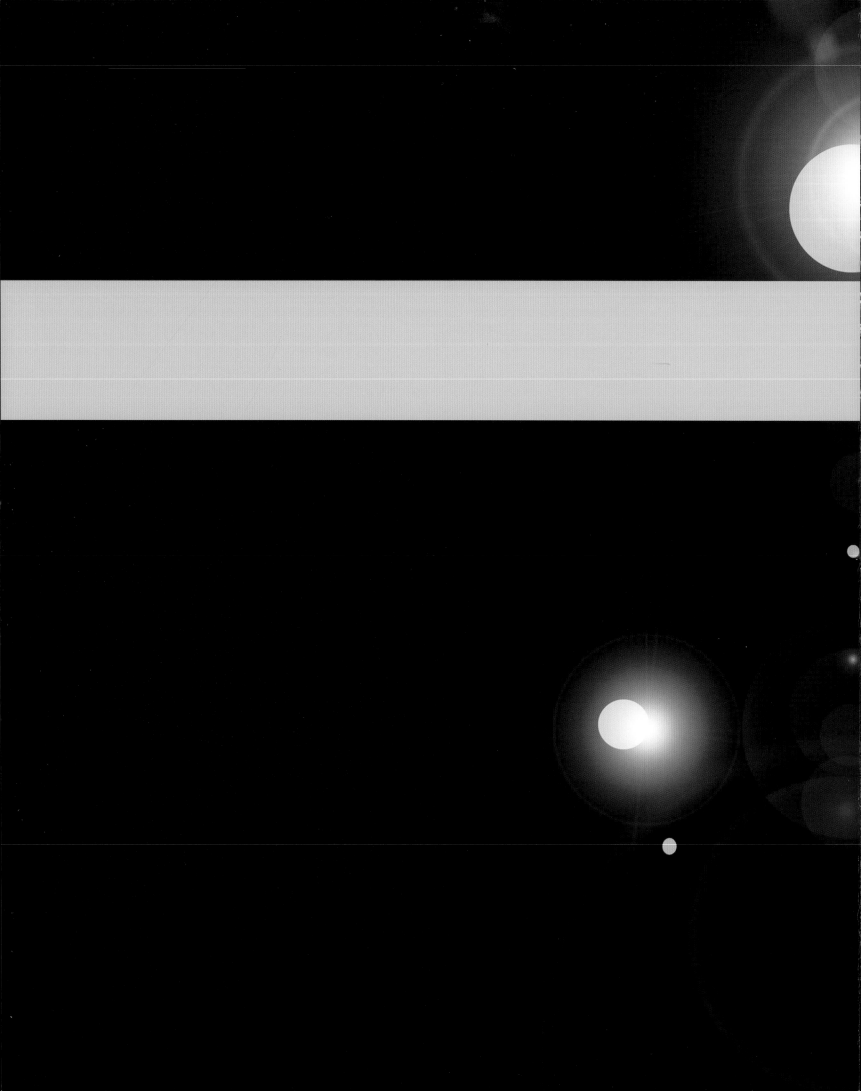

CHAPTER 01

Tom D. Crouch

WHY DO AIRPLANES LOOK THE WAY THEY DO?

Sir George Cayley (1773–1857) called his invention a "flying parachute." Built in 1804, the predecessor of all fixed-wing flying machines consisted of a simple pole, 4 feet (122 cm) long, with a few surfaces attached. The wing was a kite mounted on top of the pole at a six-degree angle of attack. A cruciform tail could be angled in any direction. A small weight was positioned near the nose to balance the craft. "It was very pretty to see it sail down a steep hill," remarked the inventor. "It gave the idea that a larger instrument would be a better and safer conveyance down the Alps than ever the sure-footed mule …"[1] (fig. 17).

The flight of that first hand-launched glider sweeping down an English hillside in 1804 marks the real beginning of the invention of the airplane. Cayley, a Yorkshire baronet, not only built the first gliders but also occasionally built them large enough to carry human beings on short journeys through the air. He was also the first man to understand what an airplane ought to look like, in general terms.

WHY DO AIRPLANES LOOK THE WAY THEY DO?

The answer to that question begins with the realization that weight is the great enemy of flight. A flying machine is the only craft that has to lift into the air its own weight and that of a crew and useful payload before it can even begin to do its job. The most basic goal of the aeronautical engineer is to provide a flying machine that will meet its performance requirements and offer adequate strength and safety with minimum weight. As a consequence of that unforgiving problem, a successful aircraft design will reflect the underlying technology. Cayley's first glider had a simple fixed wing, control surfaces, and a means of tying these essential elements together. Anything else would have been excess weight.

The means of achieving flight have evolved over time. Improved methods of propulsion, breakthroughs in aerodynamics, the advent of new materials, and a host of other factors that enabled us to fly higher, faster, farther, and more efficiently have reshaped our vision of the ideal airplane at critical moments during the century-long history of aviation.

EARLY PIONEERS

Let us start at the beginning. Why did the 1903 Wright Flyer, the world's first airplane, look the way it did (fig. 18)? The Wright brothers began the design process by using simple algebraic equations, and data borrowed from their predecessors, to calculate the amount of wing surface required to lift the estimated weight of their first glider, with a pilot on board, in a particular wind speed. The calculation indicated that they would either have to have an enormous wing or would have to fly their machine into a very strong headwind.

They found the solution to their problem in a hang glider designed in 1896 by the engineer and flying-machine enthusiast Octave Chanute. In order to provide maximum wing area with a relatively short span, Chanute had broken the wing in half and produced a biplane, or "two-surface machine." Test-flown that summer on the dunes ringing the southern shore of Lake Michigan, the little glider performed well. More important, it was a superb structure. Chanute had tied the two wings together with the same sort of truss that he used in building railroad bridges. He had transformed the two wings, separated by wooden uprights and linked with wire, into a single beam. It was the first modern aircraft structure, the strength of which could be analysed using standard engineering methods.[2]

The biplane had an enormous advantage over the monoplane. In addition to providing maximum wing area in a short span, the biplane structure offered great strength and rigidity in a lightweight package. The relatively thin, wooden monoplane wing of the period was very weak, and had to be externally braced. Guy wires attached to king posts on the top and bottom of the fuselage held the wing in place. The arrangement was far from satisfactory. The "torsion tube," that portion of the wing that can resist twisting forces, was only as thick as the wing, and as large as the distance between the two wing spars. The torsion tube of a biplane wing, by contrast, was as thick as the entire area between the top of the upper wing and the bottom of the lower wing. Such a beam structure was capable of resisting enormous force.

The catastrophic failure of monoplane wings as a result of forces in torsion and compression remained a problem through World War I. Why, then, did people continue to build monoplanes? Even at the speeds that were being flown up to

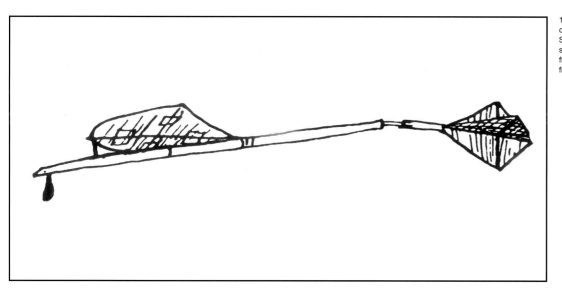

17. The initial concept and configuration for the airplane. Sir George Cayley drafted this sketch of a glider, which he later flight-tested in 1804. It was the first true airplane model.

18. Wilbur and Orville Wright made the first powered, sustained, and controlled heavier-than-air flight at Kitty Hawk, NC, on December 17, 1903.

1914, monoplanes produced marginally less drag and more lift per square foot, because of the aerodynamic interference between the two wings of a biplane. That efficiency adds up to a little more speed. In the era of fierce aeronautical competition, a few extra miles an hour seemed worth the slight extra danger represented by a weaker wing structure.

Engineer Louis Bechereau made monoplanes even faster by introducing the notion of streamlining in 1912–13. His Deperdussin racers, the fastest things in the sky, were externally braced monoplanes with a monocoque (single shell) fuselage. The strong, lightweight monocoque design was constructed of thin strips of tulip wood, criss-crossed and glued together in three layers inside a mold, then covered, inside and out, with varnished fabric. The resulting shell was an incredibly light and very strong structure capable of carrying the stresses of flight without the need for heavy internal bracing.

During World War I, Anthony Fokker and other designer/manufacturers began to build wings that were thick enough to contain a strong internal structure. The result was a cantilevered wing that did not require external supports. It was the solution to the problem of monoplane wing collapse.

As more powerful engines led to higher speeds, streamlining and drag reduction became important concerns. With the faster speeds and improved structures, the biplane gave way to the more efficient monoplane. By the late 1920s, the Lockheed Vega was every youngster's notion of what an airplane ought to look like. A sleek, high-wing monoplane with a monocoque fuselage and a thick cantilevered wing, powered by a radial engine hidden inside a drag-reducing cowling, the Vega summarized all the lessons of aircraft design learned to date.

New Forms and Materials

The Vega had taken shape on the drawing board of John Knudsen "Jack" Northrop, one of the most influential designers in the history of aviation (fig. 19). Always looking toward the future, Northrop had turned his attention to the design of all-metal airplanes by the end of the 1920s. The story of metal aircraft construction had begun in Germany before World War I. Metal aircraft had

flown in combat during the War, and served as airliners during the immediate post-war period. For the most part, however, metal had simply been substituted for wood without taking full advantage of the new material.

Wood remained a perfectly acceptable material for aircraft construction. The De Havilland Mosquito, one of the fastest, highest-flying aircraft of World War II, was built almost entirely of wood. Why, then, did the general switch from wood to metal take place? The simplest answer is that metal is tougher, more uniform, and can be used to create stronger, lighter structures. It can be machined with precision for standardized production. The development of metallurgical processes that solved the problem of aluminum corrosion set the stage for the era of all-metal aircraft.

Jack Northrop was one of those who developed new structural forms that took full advantage of the potential of metal. His Alpha (1930) and Gamma (1933) were sleek, low-winged monoplanes with a monocoque fuselage, and stressed-skin, cantilevered wings (fig. 20). A stressed skin carries a significant portion of the aerodynamic load. Northrop developed an innovative "multicellular" wing in which a series of longitudinal spars intersected with the ribs, creating multiple "cells" that gave the wing its strength.

Yet if these aircraft pointed to the future, they also carried a great deal of the past with them. Both the Alpha and Gamma, for example, were open-cockpit aircraft with fixed landing gear. Unwilling to break into his multicellular wing to house retractable landing gear, Northrop fitted streamlined fairings to his fixed gear.

A generation earlier, airplanes had flown so slowly that the increased drag of fixed landing gear was a matter of little consequence. In a fully streamlined, low-drag, all-metal craft like the Alpha, however, even the streamlined fixed landing gear represented a significant proportion of total drag.[3]

Such problems notwithstanding, the early Northrop aircraft represented a major step toward the first generation of modern airliners. Northrop himself played a key role in the design of the wing of the Douglas DC-1, DC-2, and DC-3 aircraft,

19. The Lockheed Vega, designed by Jack Northrop in 1929. The Lockheed Vega was one of the first production aircraft to be fitted with a drag-reducing NACA cowling.

20. The Northrop Alpha, built in 1930. Jack Northrop's Alpha made full use of the potential of new materials.

which first flew between 1933 and 1935. There was nothing transitional about these machines. With fully retractable landing gear, controllable pitch propellers, modern cockpit instruments, a fully realized metal structure, and flaps that could be extended to improve low-speed handling characteristics at the critical moments of take-off and landing, the DC-3 was one of the most important designs in the history of flight, and set the pattern for the World War II generation of aircraft.

THE JET AGE

The advent of the jet engine, independently invented by the Englishman Frank Whittle and the German Hans von Ohain immediately before World War II, was one of the great turning points in the history of flight. The speed of these vehicles created new problems for aerodynamicists and aircraft designers, however. As an aircraft approached the speed of sound, the sudden increase in drag was so alarming that some researchers spoke of a sound barrier. The shockwave generated as the machine approached the speed of sound disrupted the flow of air around the wing, leading to increased drag and control problems.

As early as 1935, even before the appearance of the first jets, the German aerodynamicist Adolf Busemann had forecast these problems, and suggested that designers could avoid them through the use of swept wings. During World War II, German engineers made use of Busemann's work, and that of fellow engineer Alexander Lippisch, to design the first generation of jet- and rocket-powered combat aircraft and even more advanced proposals for swept-wing airplanes that would never leave the drawing board. In the post-war years, engineers on both sides of the Iron Curtain drew on the German research to produce designs that would once again change our notion of what a modern airplane ought to look like, and, as historian Edward Constant has reminded us, what an airplane sounds like.[4] For people of my generation, who were youngsters during the Korean War, the idea of swept-wing Sabre jets jousting with MiGs in the high thin air over the Yalu River still makes the heart race.

Even the look of the streamlined, swept-wing fighters, bombers, and jetliners has changed over time. In 1952, the prototype of the delta-winged

21. The initial configuration of the F-102 jet fighter aircraft, 1952. This prototype proved incapable of supersonic flight.

22. The "wasp-waisted" version of the F-102 aircraft after implementation of Richard Whitcomb's "area rule." This redesign enabled the F-102 to break the sound barrier.

Convair F-102 jet fighter was having surprising difficulty flying faster than sound (fig. 21). It occurred to Richard Whitcomb, a brilliant engineer employed at the Langley Laboratory of the National Advisory Committee for Aeronautics (NACA), that creating a "wasp-waisted" fuselage in the area where the wings meet the body of the aircraft would produce a total cross-sectional area of fuselage, wings, and tail that represented an ideal streamlined body. Dubbed the "area rule," it was counter-intuitive, but repeated wind tunnel tests suggested that it was true. Flight tests of the wasp-waisted F-102 proved it (fig. 22).

Whitcomb launched another era of change in 1966, when he announced the design of a "supercritical wing." Intended drastically to reduce the drag on a wing approaching the speed of sound, the airfoils that emerged from the Langley wind tunnels looked very odd, but they worked. Whitcomb invented the winglet, as well. Upturned surfaces on the wingtips of an airliner, winglets produced additional thrust by harnessing the energy of the air rushing around the outer edge of the wing, resulting in improved fuel economy and increased range.[5]

Other changes in appearance came with increases in speed. The traditional cambered airfoil of the subsonic airplane gave way to very different shapes that provided lift for machines flying at twice the speed of sound. Experimental supersonic bombers and airliners shared delta-shaped wings and noses that dropped down for take-off and landing and rose into a streamlined position for normal high-speed flight. The look reflected aerodynamic realities.

CHANGING PRIORITIES

A host of factors, from the rise of effective anti-aircraft missiles to environmental concerns and the rising cost and complexity of very high-speed aircraft, led to a redefinition of progress that emphasized factors other than the ability to fly very fast and very high. Once again, these requirements were reflected in the changing appearance of the airplane. The era of the wide-body airliner, inaugurated by the Boeing 747, was the result of the realization that the ability to carry more people and cargo over long distance at high subsonic speeds would prove more profitable than supersonic transport. The wide-body shape resulted from the need to carry two standard cargo containers side by side.

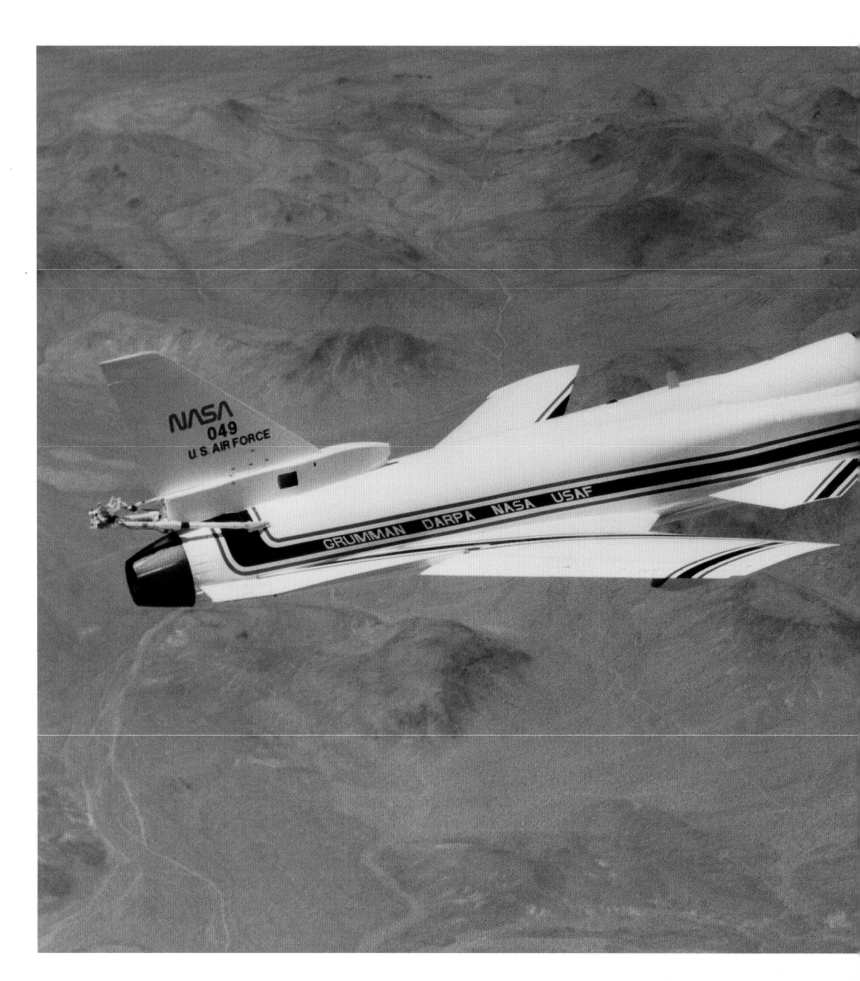

23. A joint NASA–US Air Force flight-test program studied the feasibility of forward-swept wings for greater performance at high angles of attack. The X-29 aircraft first flew in 1989.

24. With its diamond shape,
the F-117A Stealth Fighter was
designed to escape detection
by radar. In the computer age,
aerodynamic considerations
could take second place to
other characteristics.

In similar fashion, military reality underscored the fact that the ability to fly faster and higher was less important than other qualities, such as the ability to avoid detection by radar or infrared sensors. Stealth technology began in 1966, with a paper published by the Russian mathematician Pyotr Ufimtsev, who developed equations that enabled an engineer to calculate the radar signature of any design. The calculations were so complex and the problems so daunting, however, that this approach was impractical until the rise of a new generation of powerful computers and software. In any event, computer-operated "fly-by-wire" systems would be essential in order to control the otherwise unflyable shapes that would be invisible to radar. As with the forward-swept wing of the Grumman X-29, the first-generation stealth aircraft were so aerodynamically unforgiving that they could not be flown without the constant intervention of a computer (fig. 23).

In 1975, Denys Overholser, a radar specialist at the Skunk Works, the famous Lockheed advanced-projects office, developed a computer program that could predict the radar cross-section of a given design. A model of the resulting optimum shape, a flat faceted object that engineers took to calling the "Hopeless Diamond," was tested in an indoor radar chamber. The final product of the research, the F-117 Nighthawk, reached operational units in 1982 (fig. 24).[6]

The F-117 was a wedge, all planes and facets designed to reflect radar energy anywhere but back to the receiver. Northrop engineers, using more advanced computer programs, took a very different and more sophisticated approach to the design of the B-2, a stealth bomber with graceful flowing lines. Cloaked in tight security during its development, the airplane was publicly unveiled on November 22, 1988. A range of new technologies had converged to produce a genuine flying wing.

A generation before, Jack Northrop had struggled to develop a flying wing bomber. The limitations of technology had barred the way to complete success. Half a century later, on-board computers solved those problems. At the end of the century, as at the beginning, evolving technology was still changing the shape of the airplane.

Notes

1. Sir George Cayley, quoted in Charles H. Gibbs-Smith, *Sir George Cayley's Aeronautics, 1796–1855*, London (HMSO) 1962, p. 18.

2. For information on the early history of flight technology, see Charles H. Gibbs-Smith, *Aviation: An Historical Survey From Its Origins to the End of World War II*, London (HMSO) 1970. For more on Octave Chanute, see Tom D. Crouch, *A Dream of Wings: Americans and the Airplane, 1875–1905*, New York (W.W. Norton) 1981.

3. Walter Vincenti, "The Retractable Airplane Landing Gear and the Northrop 'Anomaly': Variation-Selection and the Shaping of Technology," *Technology and Culture*, vol. 35, 1994, pp. 1–33.

4. Edward W. Constant, II, *The Origins of the Turbojet Revolution*, Baltimore (The Johns Hopkins University Press) 1980.

5. John D. Anderson, Jr., *A History of Aerodynamics and Its Impact on Flying Machines*, Cambridge, UK (Cambridge University Press) 1997, describes the work of Richard Whitcomb and is far and away the best history of the subject.

6. Herbert Fenster, *The $5 Billion Dollar Misunderstanding*, Annapolis, Md. (Naval Institute Press) 2001, offers insight into the history of stealth technology.

CHAPTER 02

Dominick A. Pisano

THE AIRPLANE AND THE STREAMLINE
IDIOM IN THE UNITED STATES

As a means of travel, commerce, and warfare, the airplane has brought about significant changes to the lives of people throughout the world. Indeed, its importance is so great that it can be said to have changed the course of civilization. Less obvious, however, and greatly overshadowed by its commercial and war-making potential, is the airplane's effect on American aesthetic and cultural values in the period beginning immediately before World War I and ending in the late 1930s.

CULTURAL SYMBOLISM

The airplane's influence as a cultural symbol can be said to have begun in Italy before World War I, when an artistic movement called Futurism—led by Filippo Tommaso Marinetti and later carried forward by his successors—emerged in reaction to nineteenth-century ideals of beauty.

The Futurists celebrated the dawn of the new technological age with peans to the raw speed and power of the newest technological marvel, the airplane. This they viewed as an object to be revered for its purity of form and its significance to the newly arrived twentieth century. Later on in Europe, the theories and work of such visionaries as Le Corbusier and Erich Mendelsohn suggested that the airplane's functional and expressionistic meaning could be encapsulated in the architectural media of concrete and steel. Finally, through the cumulative influence of these men, the symbol of the airplane arrived in the US. Here it was exploited by a new breed of artist-businessman, the industrial designer (fig. 25), and was transformed into a glamorous and profitable manifestation of a future age of speed and efficiency. Amidst universal feelings of economic hopelessness, this peculiarly American vision of the airplane was forged.

With the advent of mass-produced, commercially available, industrially designed products that were advertised and packaged to appeal to the consumer's desire for glamour and romance in everyday surroundings, the stage was set for the airplane to become the most important cultural symbol of the period. As industrial designers began to gain wide acceptance by the mid-1930s, the symbolic import of the airplane began to manifest itself in a way that was to have an overwhelming effect on the cultural attitudes of millions of Americans.[1]

After a brief love affair with another important cultural symbol—the skyscraper—during the 1920s, designers in the following decade turned their attention to machines of transport. Much of the credit for this new trend must go to two of Le Corbusier's published works: *Towards a New Architecture* (1927) and *Aircraft* (1935), which celebrated the rational form of the ocean liner, automobile, and airplane and promoted them as sources of modern architectural inspiration. On Le Corbusier's influence, Jeffrey Meikle has commented:

Despite a functionalist reputation, Le Corbusier himself steered American designers toward an expressionist machine aesthetic, according to which consumer products, transportation machines, buildings, and interiors would reflect attributes of the machine—speed, power, precision, machined surfaces, and impersonality.[2]

By the late 1930s, the sleek figure of the Douglas DC-3 (fig. 26), the most striking example of streamlined aircraft, had begun to dominate the air lanes. Despite commercial flying's exclusivity during this time, the sight of the streamlined DC-3, shiny and metallic, with its distinctive hemispherical nose section, tapering cylindrical fuselage, and curvilinear construction, was becoming familiar to people everywhere. The DC-3 boasted several important technological advances, but its looks also had an impact: the DC-3 suggested speed and motion. Thus, it was only natural that industrial designers looked to its clean lines as a way of identifying with the public's notion of speed and glamor, and perhaps more importantly, as a means of selling their products.

By 1936, Sheldon Cheney, an enthusiastic proponent of streamlining, was singing the praises of the streamline idiom in industrial art and design and of the airplane as "the most conspicuous object of the new age." "We live in a world of streamlined vehicles," Cheney wrote:

The streamline as a scientific fact is embodied in the airplane. As an aesthetic style mark, and a symbol of twentieth-century machine-age speed, precision, and efficiency, it has been borrowed from the airplane and made to compel the eye anew, with the same flash-and-gleam beauty re-embodied in all travel and transportation machines intended for fast going.[3]

25. Raymond Loewy, arguably the most prominent industrial designer of the twentieth century, typified his profession. Here he poses in the mocked-up designer's office that he created with Lee Simonson for the Metropolitan Museum, New York, in 1934. Loewy's clients included the Pennsylvania Railroad, Coca-Cola, Studebaker, and the American Tobacco Company. In 1937 Loewy designed the S-1 locomotive for the Pennsylvania Railroad (see fig. 28). After World War II he worked on making NASA's Skylab experimental space station habitable for long periods in space.

STREAMLINING AND FUNCTIONALITY

Functional aerodynamic streamlining had been a long time in arriving, however. Leonardo da Vinci may have been the first man in recorded history to understand the principle of streamlining, but he applied his knowledge to the motion of a body traveling through water rather than one moving through air. It was not until the early part of the nineteenth century that Sir George Cayley, often called the father of modern aeronautics, mentioned streamlined bodies in connection with airships and used the term "solid of least resistance." In his paper "On Aerial Navigation," published in 1810, Cayley spoke of the resistance of air to a moving body, but his conclusion was that "the whole of this subject is of so dark a nature as to be more usefully investigated by experiment than by reasoning."[4]

Nevertheless, Cayley's drawings of a solid of least resistance, based on the shape of a trout and a dolphin, are a good indication that he understood the basic principle of drag reduction. This was the idea that the flow of a steady stream of air around a solid was essential to the propulsion of that solid through the medium of air. It is no accident that the shapes represented by Cayley as the solid of least resistance bear an unmistakable resemblance to the teardrop shape—

characteristic of penetration—that was so prevalent in industrial design during the 1930s.

Cayley's theories of a streamlined body moving through air were apparently lost on succeeding generations of aeronautical experimenters, however, judging by the clumsy box-like constructions that dominated the early stages of aircraft design. Nevertheless, in the early part of the twentieth century other aerodynamic theorists, such as F.W. Lanchester and Ludwig Prandtl, took up the cause of streamlining as a means of eliminating the parasitic drag that had impeded the effectiveness of early aircraft.

In 1907, Lanchester had called attention to the problem of air resistance caused by drag. According to Ronald Miller and David Sawers, he theorized that "the drag of a perfectly streamlined airplane should be no more than that caused by the friction of the air passing over its surface and that needed to sustain it in the air."[5] Yet few designers appreciated the significance of Lanchester's work. Similarly, Ludwig Prandtl, who later became director of the renowned aeronautical research institute at Göttingen, and who agreed in theory with Lanchester, argued that the flow of air around a solid body would either be turbulent or smooth ("laminar") and that drag would be reduced markedly if laminar flow

26. Considered to be the first modern airliner, the Douglas DC-3 incorporated design features such as variable-pitch propellers, wing flaps, air-cooled radial engines, engine cowlings that reduced drag, and an aerodynamically improved (streamlined) unbraced cantilever monoplane airframe. Aside from its technological innovations, its clean and aesthetically pleasing appearance inspired the industrial designers of the 1930s.

was sustained. Early on, however, considerations of weight rather than shape seemed to play a more important role in aircraft design and the resulting aircraft of the period before World War I seem to bear this out.[6]

In 1929, however, when Professor B. Melvill Jones read his paper "The Streamlined Aeroplane" before the members of the Royal Aeronautical Society in London, aeronautical engineers began to understand the reasons for streamlining and took steps to evolve airplane shape in an organic and functional way. Still, the externally braced, box-girder biplane configuration developed by the Wright brothers remained the norm until the 1920s.

Attempts to overcome the aerodynamic problems of the wooden, box-girder biplane with its external bracing wires, fixed undercarriage, and exposed engines were made in Germany by Hugo Junkers and Adolph Rohrbach and in England by Oswald Short. Junkers, the well-known German aircraft developer, had taken the first steps in the direction of practical streamlined aircraft with the

27. A spin tunnel model of a Stearman 85, c. 1937. The Stearman Company produced a variety of biplanes that culminated in their famous Kaydet trainer of 1934. The model shown here represents an unsuccessful experimental prototype, designated XOSS-1, for a US Navy reconnaissance plane. (Stearman lost this commission to Chance Vought's famous OS2U Kingfisher.) The wooden model features the streamlined NACA cowling surrounding the engine; the cowling's curved shape enabled aircraft to fly more efficiently by producing less resistance, or drag, as they flew. Perhaps the most famous early airplane to use the NACA cowling was the Boeing 247 airliner of 1933, which was showcased at the Century of Progress Exposition in Chicago that year.

introduction in 1915 of an all-metal, internally braced, low-wing cantilever monoplane, the outer covering of which was made from corrugated sheet-metal. Junkers's innovations, however, ultimately did not solve the problem of aerodynamic drag because the corrugated metal skin, which was characteristic of a long line of Junkers aircraft, was found to be drag-inducing. A few years later, Adolph Rohrbach introduced the concept of the smooth stressed-skin wing structure—further refined by John Knudsen "Jack" Northrop in the US—in which a substantial part of the load is carried by the skin. And in 1920, Oswald Short's all-metal, stressed-skin biplane, the Silver Streak, made its appearance in Great Britain. Although technically advanced for its day, its design was not widely accepted by the conservative British aeronautical community.

It was not until the 1930s in the United States that the first practical streamlined aircraft began to make an appearance. With the introduction of the Boeing Model 200 Monomail (1930) and Boeing Model 246 (Y1B-9A, 1932) and the Northrop Alpha (1930), the streamline revolution in aircraft design had begun. Although it was never placed in production, the Monomail was a highly sophisticated all-metal aircraft with smooth skin, low-wing monoplane construction and other important streamline developments such as a cowled engine and semi-retractable landing gear. From the design of the Monomail came the first significant streamlined airliner, the Boeing 247, which was overshadowed by the Douglas DC-series transports. The Northrop Alpha, designed by Jack Northrop in 1929, featured all of the streamlining innovations of the Monomail with the exception of retractable gear. Along with the Boeing 247, the Alpha was a prime source of inspiration for the first commercially practical streamlined aircraft, the Douglas DC-3.[7]

With the introduction of the Douglas DC-3 in 1936, functional aerodynamic streamlining had been accomplished. More importantly, though, Americans now had an important cultural symbol for the age. Commenting on the effect of the streamlined aircraft on American society, Donald Bush has remarked aptly that "the logic of aerodynamics was now clearly stated in a form that revealed its function while summing up and symbolizing the ideas of flight, lift and low resistance."[8] As we shall see, these ideas were

to play an important role in American culture as a result of their transference in the form of the streamline idiom to the everyday lives of great numbers of Americans.

Even before the development of the fully streamlined and economically viable Douglas DC-3, an event had taken place that fired the imagination of the American public and united the country's feelings toward the airplane and aviation. That event, of course, was Charles A. Lindbergh's solo flight across the Atlantic in 1927. Lindbergh's flight, the financial boom in aviation that took place in 1928–29, and the federal regulation of aviation that preceded these events in 1926, were all catalysts in the creation of viable commercial air travel in the United States. Subsequently, the public's awareness of the airplane's commercial potential increased. These developments, along with the technological improvements to and streamlining of aircraft, which increased their speed and efficiency, made it economically possible for air travel to become widely accepted in the United States.

RISE OF THE INDUSTRIAL DESIGNER

It was little wonder that the airplane now began to fulfill Sheldon Cheney's billing as "the most conspicuous object of the new age." The impetus for the streamline movement, however, came not only from people such as Cheney, who saw it as an emblem of the spirit of the age, but more importantly from a small group of industrial designers, based in New York, who were intent on taking advantage of its commercial possibilities. Led by such proponents as Walter Dorwin Teague, Raymond Loewy, Henry Dreyfuss, and Norman Bel Geddes, the streamline movement in industrial design began to occupy a prominent place in American life.

Unlike Loewy, Dreyfuss, and Bel Geddes, Walter Dorwin Teague was interested more in the psychological benefits of streamlining than in its practical applications to land vehicles or aircraft. As an example of the pleasurable aesthetic connotations of modern aircraft, Teague not unnaturally cited "the constant ratios of proportion" and "the quality of line which we find most highly developed … in a Douglas transport plane, where you see the same type of form repeated in the engine and in the fuselage, in the wings and the tail—the same line recurring again and again; that long line with a sharp parabolic curve at the end, which we have come into the habit of calling 'streamline.'"[9] According to Jeffrey Meikle, "Teague based the validity of streamlining for all objects, whether they moved or not, on its ability to express the spirit of the age."[10]

Raymond Loewy and Henry Dreyfuss went a step further than Teague and promulgated the idea that streamlining was not only pleasing to the eye but also functionally appropriate for such earth-bound vehicles as the locomotive and automobile. In 1936, Loewy and the engineering staff of the Pennsylvania Railroad designed the K4S streamlined shroud for the steam locomotive that pulled the famous *Broadway Limited* train. In 1938, Loewy collaborated on the design of the S-1 locomotive (fig. 28), a sleek, horizontal-lined machine reputed to be the "largest and fastest high-speed steam engine ever to be placed in service in this country."[11] Dreyfuss worked for the Pennsylvania Railroad's chief competitor, the New York Central, in designing the J-3A locomotives (fig. 29), which hauled the *Twentieth Century Limited* from New York to Chicago. These, however, were after-the-fact imitations of the Union Pacific Railroad's truly innovative, functionally streamlined *City of Salina* (1934) and the Chicago, Burlington & Quincy Railroad's *Zephyr* (1934). Both were powered by diesel-electric locomotives, were wind tunnel tested, and, like modern aircraft, were designed with monocoque construction (*i.e.*, a hollow structure without internal bracing in which most of the stresses are carried by the skin).[12] William Stout, aircraft and train designer, called them "wingless aircraft on tracks."[13]

Norman Bel Geddes, the doyen of the streamline adherents in the United States, published an article that appeared in *Atlantic Monthly* in 1934 outlining in layman's terms the scientific ramifications of streamlining.[14] Geddes concluded that although, as yet, not enough was known about the effects of streamlining on land vehicles and ocean-going vessels, the possibilities for increased speed and efficiency in these machines as a result of the lessening of the effects of air resistance were well worth exploring. Geddes' book *Horizons*, published in 1932, contained a wealth of photographs and illustrations that depicted streamlined vehicles and concepts for transportation machines. Included among these

was an immense (though theoretical) transatlantic *Air Liner Number 4*, which Geddes had designed with the help of the German aeronautical engineer Otto Koller. Utilizing twenty engines, together capable of creating 38,000 horsepower, *Air Liner Number 4* would be capable of cruising at 100 miles (160 km) per hour and of flying between Chicago and Plymouth, England, in forty-two hours.[15]

In 1933, the Chrysler Corporation undertook the design of the first truly mass-produced streamlined automobile, the Chrysler Airflow, under the leadership of Carl Breer. The Airflow had a welded, trussed box-frame construction (designed by Alexander Klemin, head of the Daniel Guggenheim School of Aeronautics at New York University) in which girders and body panels were integrated into a shallow frame, yielding a highly rigid but sturdy structure. The Airflow's style grew out of hundreds of wind-tunnel tests that were completed on models of the automobile by the design team in order to reduce drag and noise and to improve stability, and was promoted in its advertising as growing out of its functional design.[16]

According to David Gartman, the Airflow was representative of the attempt of automotive engineers to gain precedence over the automotive stylists who had become increasingly influential; thus it was symbolic of the conflict between scientific and aesthetic standards in automobile design.[17] Ultimately, the stylists won out. Although the Airflow sold well in 1934, its first year of production, by 1937 sales had begun to lag and the car was taken off the market. The Depression undoubtedly contributed to its demise, but

28. (above) Created by Raymond Loewy in 1937, this rendering of the Pennsylvania Railroad's S-1 streamlined locomotive is obviously intended to convey the machine's sleekness and speed. The S-1 was exhibited at the New York World's Fair in 1939.

29. Henry Dreyfuss designed the streamlined J-3A locomotive pictured here for the New York Central Railroad (NYC) in 1938. The J-3A drew NYC's famous passenger train the *Twentieth Century Limited*, which ran between New York and Chicago.

according to Donald Bush, Raymond Dietrich, head of exterior design at Chrysler at the time, felt that the Airflow "failed to elicit a response from buyers (despite its claims) because it did not have a profile that implied forward motion."[18] Buyers remained skeptical of the technical claims made for streamlined autos and were attracted rather by the appearance of streamlining in such stylish automobiles as the 1936 Lincoln Zephyr and Cord 810, and the 1940 Lincoln Continental.

SPREAD OF THE STREAMLINE STYLE

Culturally, the streamline idiom was to have more far-reaching effects on what has been called by David Gebhard the "commercial vernacular" style in American architecture. Gebhard, a leading critic of the Streamline Moderne style in architecture, has remarked that historians and critics in the field have tended to agree that the functional International Style, which developed in Europe (in the Bauhaus in particular), was the most important influence on modern American architecture. International Style theorists had seen the machine as a static object, and the box-like, cubistic designs they produced tended to bear out their views. "Expressionist" architects such as Erich Mendelsohn, by contrast, saw the machine as a driving, dynamic force, one that was characterized by machines of transport. Influenced by Mendelsohn and the industrial designers, commercially successful architects in the United States began to borrow their predilection toward speed and motion and incorporate it into their designs. The Streamline Moderne style in American commercial vernacular architecture had been born. And in the 1930s especially, when the airplane began to gain wide acceptance in the popular consciousness as the transportation machine of the future, the Streamline Moderne style became the prevailing architectural idiom in the United States.[19]

The showplace of the Streamline Moderne style in American architecture, the New York World's Fair held in 1939, symbolized the overwhelming influence of both the industrial designers and the airplane on what was considered to be the architecture of the future. Norman Bel Geddes' Highways and Horizons Exhibit built for General Motors (figs. 30, 31), for example, with its buildings characterized by rounded forms and curvilinear motifs, was easily the most integrally streamlined and futuristic exhibit of the fair. And

30, 31. Two artists' representations of the streamlined General Motors Building, called the General Motors Highways and Horizons Exhibit, or "Futurama," designed by Norman Bel Geddes for the 1939 New York World's Fair. In addition to the building's modern appearance, the exhibit inside projected a theme of highway progress and trends in motor transportation facilities of the future. Visitors, in moving sound-chairs, toured a vast miniature USA as it might appear in 1960.

although it was more conventionally streamlined, Raymond Loewy's Chrysler Motors Building, with its "ovoid center section flanked by slender pylons with streamlined chevrons," was a marked example of the idiom.[20]

By the end of the decade, the streamline style had become all-pervasive. Even common household objects and consumer products such as the thermostat, pencil sharpener (fig. 32), refrigerator, iron, vacuum cleaner, electric fan, and toaster had succumbed not only to the industrial designer's whims but also to the penchant in American society for the streamlined form. Nevertheless, despite the overwhelming popular acceptance of

the streamline idiom, and of the airplane that had made it possible, the style that had characterized the period of the 1930s was by 1940 fading into history. If the symbolic culmination of the streamline style in American architecture had been the structures erected for the New York World's Fair, then the destruction of these buildings was perhaps symbolic of the end of the streamline era in American life. The promise of the technological paradise conjured up by the World's Fair never completely materialized. Although its directors had touted it at great length as benefiting the man in the street, concerns of world survival had by 1939 begun to take precedence over the promise of a streamlined future.

As Europe prepared for war, the airplane and the expectation of a future befitting its shiny metallic form began to give way to visions of a machine with a sinister potential for destruction unmatched in the annals of modern history. By the time the United States had entered World War II, the streamlined era, with its utopian vision of a future transformed by technology, had ended.

32. Raymond Loewy's striking streamlined pencil sharpener, patented in 1934 but never marketed, symbolizes the extremes to which industrial designers went to sell products by imputing motion to stationary objects.

33. A wind tunnel model of the Northrop P-61 Black Widow. The beautifully streamlined Black Widow was the first American plane to be specifically designed for night operations, first flying into combat in 1944. It had a top speed of 425 miles (684 km) per hour at a maximum altitude of 46,000 feet (14,020 meters), being powered by two 2100 horsepower engines. Northrop built some 700 examples of the plane. They mostly operated as night fighters over the Pacific, their noses housing radar to assist in intercepting enemy bombers.

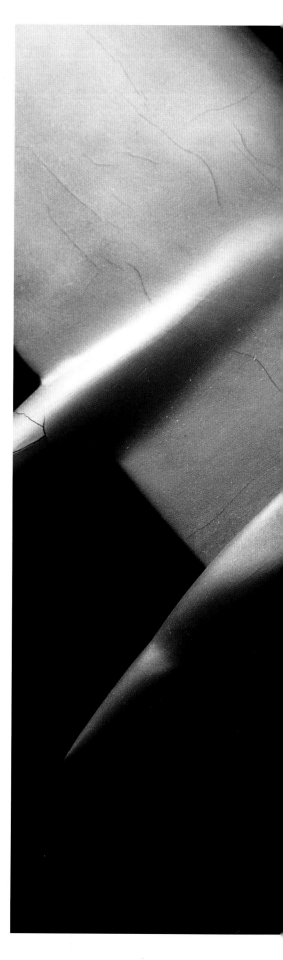

Notes

1. One of the most significant events in Europe to affect the attitudes of American art critics and designers toward the machine in art was the Exposition Internationale des Arts Décoratifs et Industriels held in Paris in 1925. The Paris show was viewed as the harbinger of a new wave in aesthetics, and it profoundly influenced the attitudes of a few important members of the relatively new profession of industrial design in the US. See Jeffrey L. Meikle, *Twentieth Century Limited: Industrial Design in America, 1925–1939*, 2nd edn., Philadelphia (Temple University Press) 2001, pp. 24–29.

2. *Ibid.*, pp. 30–31.

3. Sheldon Cheney and Martha Cheney, *Art and the Machine*, New York (Whittlesey House) 1936, pp. 16 and 97–98.

4. Sir George Cayley, quoted in Charles H. Gibbs-Smith, *Sir George Cayley's Aeronautics, 1796–1855*, London (HMSO) 1962, p. 41. The discussion on aircraft streamlining has been influenced by the following: Charles H. Gibbs-Smith, *Aviation: An Historical Survey from its Origins to the End of World War II*, London (HMSO) 1970, p. 9; Charles H. Gibbs-Smith, *Sir George Cayley's Aeronautics, 1796–1855*, p. 41; Ronald Miller and David Sawers, *The Technical Development of Modern Aviation*, London (Routledge & Kegan Paul) 1968, pp. 53–71; Noel Pemberton-Billing, *The Aeroplane of Tomorrow*, London (Robert Hale Ltd.) 1941, pp. 27–49; James Hay Stevens, *The Shape of the Aeroplane*, London (Hutchinson & Co., Ltd.) 1953, pp. 105–15; David Mondey (ed.), *The International Encyclopedia of Aviation*, New York (Crown Publishers) 1977, pp. 12–19, 76–86, and 90–95.

5. F.W. Lanchester, quoted in Miller and Sawers, *op. cit* (n. 4). p. 53.

6. *Ibid.*, p. 54.

7. The Alpha's designer, Jack Northrop, was convinced that the aircraft was the leader of the revolution in aircraft design: "I had previously experimented with a wooden monocoque [Lockheed Vega], but now I developed a smooth-skinned metal monocoque. … It was the first of its type with the thin skin of the airplane carrying the structural load. … As far as the structure is concerned that which was developed on the Alpha was really the pioneer for every airplane in the sky today." Quoted in Richard Sanders Allen, *The Northrop Story, 1929–1939*, New York (Orion Books) 1990, pp. 13–14.

8. Donald J. Bush, *The Streamlined Decade*, New York (George Braziller) 1975, p. 37.

9. Walter Dorwin Teague, quoted in Meikle, *op. cit.* (n. 1), pp. 182–83.

10. *Ibid.*, p. 183.

11. Bush, *op. cit.* (n. 8), pp. 89–90; Lois F. Brand, "Looking Backward at the Future: Raymond Loewy, An Industrial Designer," in *The Designs of Raymond Loewy*, Washington, D.C. (Smithsonian Institution Press) 1975, p. 14.

12. Bush, *op. cit.* (n. 8), pp. 62–65; Meikle, *op. cit.* (n. 1), p. 157.

13. William Stout, quoted in Meikle, *op. cit.* (n. 1), p. 160.

14. Norman Bel Geddes, "Streamlining," *Atlantic Monthly*, November 1934, pp. 553–63.

15. Norman Bel Geddes, *Horizons*, New York (Dover) 1977, pp. 111–21.

16. Bush, *op. cit.* (n. 8), pp. 118–122; James J. Flink, "The Path of Least Resistance," *Invention & Technology*, Fall 1989, p. 36; David Gartman, *Auto Opium: A Social History of American Automotive Design*, London (Routledge) 1994, p. 123.

17. Gartman, *op. cit.* (n. 16), pp. 115–17.

18. Bush, *op. cit.* (n. 8), p. 122.

19. See David Gebhard, "The Moderne in the U.S., 1920–1941," *Architectural Association Quarterly*, July 1970, pp. 4–20.

20. Meikle, *op. cit.* (n. 1), p. 196.

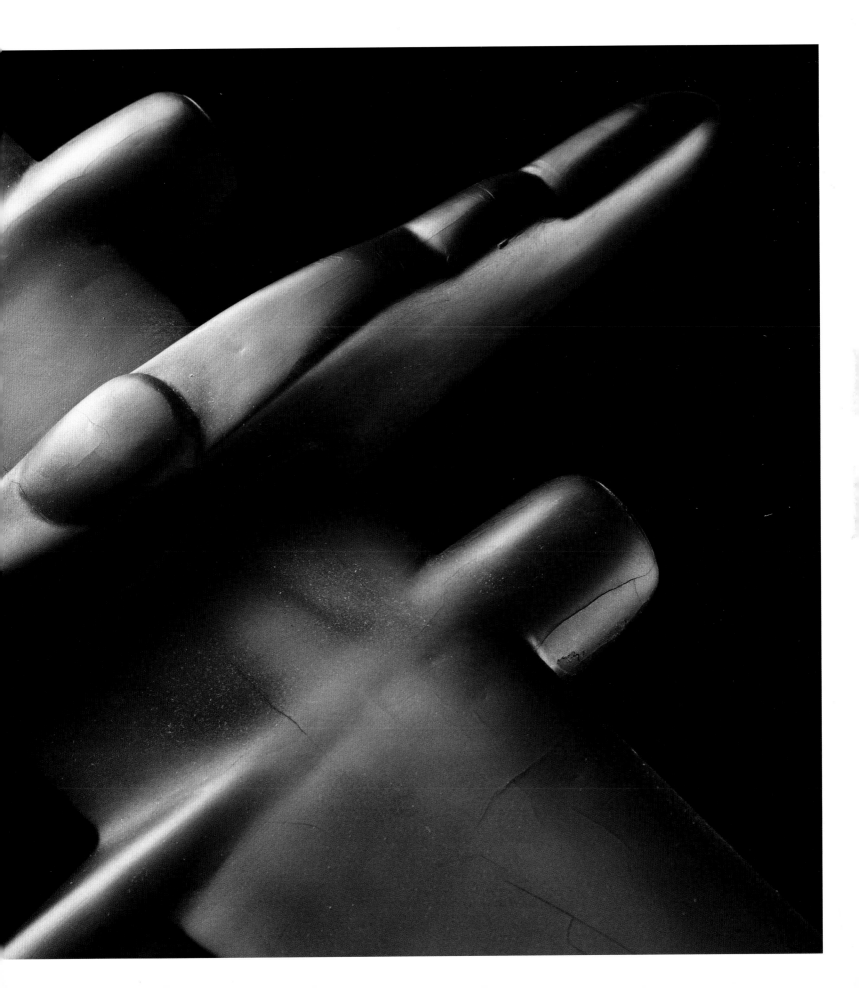

CHAPTER 03

John D. Anderson, Jr.

DESIGN FOR PERFORMANCE: THE ROLE OF AESTHETICS IN THE DEVELOPMENT OF AEROSPACE VEHICLES

34 (left). The slender, graceful Boeing 777 was the first aircraft to be designed solely on computer. Its dual high-efficiency engines render it capable of transcontinental service.

35 (below). A modified Bell X-1 spin tunnel model, *c.* 1950. The original X-1 of 1947 had straight wings and a horizontal stabilizer at its tail, whereas this painted wood spin tunnel model shows it with a swept-wing elevator. In the aftermath of World War II extensive research was completed on the concept of the swept wing for supersonic flight. Sweeping of the wings allowed the aircraft to travel faster than straight wings or surfaces. Many new configurations were studied to improve the flight characteristics of the original X-1 configuration, leading to the X-1A and X-1B aircraft.

Airplanes of all sizes and shapes fly overhead. When we pause to observe them, we see that many are aesthetically beautiful; a Boeing 777 jetliner (fig. 34), with its long, slender fuselage and gracefully swept wings, is a sight pleasing to the eye. The reason this author is an aeronautical engineer is the intense joy he experienced in building scale-model airplanes as a teenager, motivated by their aesthetic beauty. But did designers intentionally shape these airplanes to look beautiful, or was it simply a matter of form following function?

This essay explores the role of aesthetics in aerospace vehicle design, especially in cases where the designers have pushed the limits of the technology. The *American Heritage Dictionary of the English Language* defines aesthetics as "The branch of philosophy that provides a theory of the beautiful and of the fine arts." The term was first introduced in the eighteenth century by the philosopher Alexander Baumgarten in his discussions on the science of perceptible beauty. The art world deals with this term with some uncertainty. *The Oxford Companion to Art* states:

In the 20th c. there is no general agreement about the scope of philosophical aesthetics, but it is understood to be wider than the theory of fine art and to include the theory of natural beauty and non-perceptible (*e.g.*, moral or intellectual) beauty in so far as these are thought to be susceptible of philosophical or scientific study.[1]

Uncertainty notwithstanding, this passage shows that aesthetics has some connection with science, and perhaps more to the point, with the "beauty" of science.

A definition of airplane design is more concrete but has a certain degree of uncertainty as well. To quote the words of Daniel Raymer, a leading airplane designer and design educator:

Aircraft design is a separate discipline of aeronautical engineering—different from the analytical disciplines such as aerodynamics, structures, controls, and propulsion. An aircraft designer needs to be well versed in these and many other specialties, but will actually spend little time performing such analysis in all but the smallest companies. Instead, the designer's time is spent doing something called "design," creating the geometric description of a thing to be built.

To the uninitiated, "design" looks a lot like "drafting" (or in the modern world, "computer-aided drafting"). The designer's product is a drawing, and the designer spends the day hunched over a drafting table or computer terminal. However, the designer's real work is mostly mental.

If the designer is talented, there is a lot more than meets the eye on the drawing. A good aircraft design seems to miraculously glide through subsequent evaluations by specialists without major changes being required. Somehow, the landing gear fits, the fuel tanks are near the center of gravity, the structural members are simple and lightweight, the overall arrangement provides good aerodynamics, the engines install in a simple and clean fashion, and a host of similar detail seems to fall into place.

This is no accident, but rather the product of a lot of knowledge and hard work by the designer … . Design is not just the actual layout, but also the analytical process used to determine what should be designed and how the design should be modified to better meet the requirements.

Raymer goes on to say:

Those involved in design can never quite agree as to just where the design process begins. The designer thinks it starts with a new airplane concept. The sizing specialist knows that nothing can begin until an initial estimate of the weight is made. The customer, civilian or military, feels that the design begins with requirements. They are all correct. Actually, design is an iterative effort.[2]

Notice that Raymer says nothing about the role of aesthetics in the design process, but rather emphasizes the analytical and intellectual beauty of the design process.

My experience supports Raymer's definition. In college senior-level airplane design courses nothing is said about the role played by aesthetics. I can find no mention of it in any design books, old or new. Aesthetics, it would appear at first glance, does not play a role; rather, the design of an airplane appears to be based on technology and design compromises. But this may not be the whole story. In 1999 I wrote:

Airplane design is both an art and a science. In that respect it is difficult to learn by reading a book; rather it must be experienced and practiced. However, we can offer the following definition and then attempt to explain it. Airplane design is the intellectual engineering process of creating on paper (or on a computer screen) a flying machine to (1) meet certain specifications and requirements established by potential users (or as perceived by the manufacturer) and/or (2) pioneer innovative, new ideas and technology. An example of the former is the design of most commercial transports. An example of the latter is the design of the Bell X-1 (figs. 35, 36), the first airplane to exceed the speed of sound. The design process is indeed an intellectual activity, but a rather special one that is tempered by good intuition developed via experience, by attention paid to successful airplane designs that have been used in the past, and by (generally proprietary) design procedures and databases (handbooks, etc.) that are part of every airplane manufacturer.[3]

This allows that airplane design is both an art and a science, the "art" being based on the experience and subconscious of the designer— something that transcends books and design manuals. Here, I submit, is where aesthetics influences the design process. But how? This essay proposes some answers by first examining the thoughts of others on the matter. It then looks briefly at the early twentieth-century art community for interactions between the airplane and art. Finally, it examines two case histories of airplane design for examples of aestheticism.

Aesthetics and Technology

To my knowledge, the precise role of art and aesthetics in aerospace vehicle design has never been quantified. We can make a beginning, however, by noting that aerospace vehicle design is based on existing and developing technology, and although engineering technology is not, strictly speaking, applied science, we can allow that vehicle design is influenced in part by the "beauty of science." An aesthetic connection between art, beauty, and science goes back as far as the early sixteenth century, to the words of Leonardo da Vinci: "The genius of man may make various inventions, encompassing with various instruments one and the same end; but it will never discover a more beautiful, a more economical, or a more direct one than nature's, since in her inventions nothing is wanting and nothing is superfluous."[4]

Leonardo is clearly trying to explain the beauty of nature, and here we use "nature" and "science" interchangeably. The British author and physician

36. A quarter-scale model of the Bell X-1 is tested at transonic speeds in the 16 Foot Transonic Wind Tunnel at NASA's Langley Laboratory, c. 1947. Models were tested before the vehicle was flown in order to determine how the vehicle would fly.

Sir Thomas Browne reflected in 1642 that "all things are artificial, for nature is the art of God."[5] Here we have nature, hence science, equated with art, attributed to the highest authority. Mathematics is used to quantify science, and in 1917 the noted British mathematician and scientist Sir D'Arcy Wentworth Thompson commented on the combined aesthetic and useful nature of mathematics: "The perfection of mathematical beauty is such … that whatsoever is most beautiful and regular is also found to be most useful and excellent."[6] Indeed, writers and thinkers have for centuries appreciated science as having aesthetic beauty. If science, then maybe airplane design, also? Some twentieth-century intellectuals did in fact make this connection. We have from C.P. Snow, in his discussion in *The Two Cultures*:

The … line of argument draws, or attempts to draw, a clear line between pure science and technology (which is tending to become a pejorative word). This is a line that once I tried to draw myself: but, though I can still see the reasons, I shouldn't now. The more I have seen of technologists at work, the more untenable the distinction has come to look. If you actually see someone design an aircraft, you can find him going though the same experience—aesthetic, intellectual, moral—as though he was setting up an experiment in particle physics.[7]

Snow explicitly identifies airplane design as an aesthetic experience. Nevil Shute, the noted British novelist and aeronautical engineer, expressed his opinion—based on his own experience as an airplane designer—that "A beautiful aircraft is the expression of the genius of a great engineer who is also a great artist."[8] These writers are clearly making a connection between art, aesthetics, and airplane design. Finally, the noted American aviation historian Richard Smith made a direct analogy between the artist and the designer. In an essay dealing with airplane technology, he wrote:

Instead of a palette of colors, the aeronautical engineer has his own artist's palette of options. How he mixes these engineering options on his technological palette and applies them to his canvas (design) determines the performance of his airplane. When the synthesis is best it yields synergism, a result that is dramatically greater than the sum of the parts. This is hailed as "innovation." Failing this, there will result a mediocre airplane that may be good enough, or perhaps an airplane of lovely external appearance, but otherwise an iron peacock that everyone wants to forget.[9]

Along with his analogy between the artist and the airplane designer, Smith is also cautioning us that aesthetics does not guarantee a successful airplane design.

This thinking on the part of noted writers, and especially on the part of Nevil Shute, who actually designed airplanes, certainly gives the perception that aesthetics and art are an integral part of the process of airplane design. But what about the reverse process? Has airplane design had any connection with the art community? At the beginning of the twentieth century the answer was, to some extent, Yes. For example, the Russian painter Kazimir Malevich started a new approach in art that he called Suprematism—an abstract style conveying the belief that the supreme reality in the world is pure feeling, unrelated to an object. At the same time, he was influenced by an interest in modernity, and, along with many other artists from the mid-nineteenth century on, wished to capture the new images of the age. After 1908, the age included airplanes, which motivated Malevich's *Suprematist Composition: Airplane Flying* (1915), an abstract painting consisting of various blue, yellow, and red wing-like rectangles. The French painter Robert Delaunay extended the dark-toned Cubism started by Picasso and Braque to a kind of color Cubism, beginning in 1909. Also concerned with modernity, Delaunay frequently included airplanes and other aeronautical objects in his colorful paintings; indeed, what could be more in line with Cubism than the early strut-and-wire, box-like biplanes? Delaunay's *Homage to Blériot* (1914; fig. 37) is a beautiful semi-abstract painting in which one can make out a monoplane, a biplane, and a propeller. The aesthetics of the airplane had an impact on a number of other early twentieth-century artists. Airplane designers (some more than others) were creating something that was aesthetically compelling, whether they realized it or not.

All of which brings us back to the original theme, namely, to what extent does aesthetics consciously influence aerospace vehicle design, especially in regard to trade-offs in terms of technology? As discussed above, it is fairly clear

37. Robert Delaunay, *Homage to Blériot*, 1914. The painting immortalizes the achievements of the French airman Louis Blériot, who made the first flight across the English Channel in 1909. Oil on canvas, 76½ × 50½ ins (194 × 128 cm), Kunstmuseum, Basle.

that aesthetics plays little or no role in the *conscious* process of the design of an airplane or space vehicle. The design process is steeped in technology, and dictated by a myriad of design compromises. Nevertheless, the vast majority of airplanes designed over the past hundred years exhibit a degree of aesthetic beauty; some designs can easily be labeled as breathtakingly beautiful. I would suggest that there are two reasons for such beauty.

The first is related to the beauty of nature itself. The airplane designer always tries to harness nature for specific useful purposes. The laws of nature are the physical laws that govern aerodynamics, structural analysis, flight dynamics, and propulsion. These laws have an intrinsic beauty. As a result, flying machines designed to make the best use of these laws also have an intrinsic beauty. Streamlining is a case in point. Birds and fish are wonderfully streamlined because they have evolved over eons following nature's laws for minimum drag resistance. Airplanes are dictated by the same laws. We have come to appreciate birds and fish as aesthetically beautiful. Naturally, therefore, an airplane, designed by people using the same laws, will itself be aesthetically beautiful, even though the designers are not consciously trying to make it so.

Second, I believe that the difference between a good airplane designer and a great airplane designer is the extent to which his or her subconscious "feels" aesthetics. Although all designers are working with the same laws of nature, when a trade-off must be made in the design, and the scales are tipping evenly for various design choices, then I believe the great designer will subconsciously make that choice which is more aesthetically pleasing—the designer's "feelings" will make the choice. For a trivial example, consider the aspect ratio of a wing, defined for a rectangular wing as the ratio of the wing span (distance from wing tip to wing tip) to the chord (distance from the front to the back of the wing). High aspect-ratio wings are long and narrow; low aspect-ratio wings are short and stubby. If the design involves a choice between the higher aspect-ratio wing for improved aerodynamic performance and a lower aspect-ratio wing for improved structural strength, and all other aspects of the decision are essentially equal, the designer will most likely

choose the aesthetically more pleasing high aspect-ratio wing. The key proviso here is that the overall technological scales are tipping evenly for either choice; if there is even the slightest overall technical advantage to a given choice, then the rational mind of the designer takes over, and the decision will be based on technical performance, not aesthetics. After all, the designer's success and prestige are based on the performance of the vehicle rather than on its looks. But this brings us back to the first point; if the airplane is optimized using the laws of nature, it will not only perform well but it will usually have some intrinsic beauty as well.

To examine further the role of aesthetics in airplane design, let us briefly consider two design case studies, first the Douglas DC-3 airliner (fig. 38) of the 1930s, and second the Lockheed SR-71 Blackbird (figs. 41–43) of the 1960s.

THE DOUGLAS DC-3

The genesis of many airplane designs is competition. So it was in 1932, when Boeing put the final touches to the prototype of its 247 airliner, a pioneering design that Boeing felt would revolutionize commercial air travel. At that time, Boeing was a member of the United Aircraft Group, which included United Airlines. Naturally, United was first in line to receive the new Boeing airplanes, and all other airlines had to wait, putting them in an untenable competitive position. Not to be outdone, Jack Frye, vice-president of TWA (Transcontinental and Western Air, Inc. at that time) wrote a letter to Donald Douglas inviting Douglas to design for TWA a new airliner that would compete with the Boeing 247. The TWA specifications called for a trimotor—an airplane with an engine in each wing plus one mounted in the nose of the fuselage. Other trimotors existed at that time, mainly built by Fokker and by Ford. These trimotors were not particularly aesthetically pleasing. Besides, the famous Notre Dame football coach, Knute Rockne, had been killed in a crash of a Fokker trimotor in 1931, further stigmatizing this type of design. Within three days of Jack Frye's request, Douglas and his designers made the decision to design a new airplane for TWA—but it was not to be a trimotor. At a meeting on the third day, Douglas's chief engineer, James H. "Dutch" Kindelberger, stated emphatically:

I think that we're damn fools if we don't shoot for a twin-engined job instead of a trimotor. People are skeptical about the trimotors after the Rockne thing. Why build anything that even looks like a Fokker or a Ford? Both Pratt & Whitney and Wright-Aeronautical have some new engines on the test blocks that will be available by the time we're ready for them. Lots of horses ... any two of them will pull more power than any trimotor flying right now.[10]

Douglas agreed. An essential design decision was made without making a single calculation. Here is a design decision, I believe, where the subconscious minds of both Kindelberger and Douglas were influenced in part by aesthetics. A twin-engine airplane, with engines nicely blended into the leading edge of the wings, is inherently more beautiful than an airplane that also has a engine protruding from the nose. Of course, both men convinced themselves that the new engines were so powerful that two would do the job. But without definitive calculations, the decision to go with a twin-engine design had to be based on the "feelings" of these men, and aesthetics was working in their subconscious minds.

The new Douglas airplane was designated the DC-1, the forerunner of the famous lookalike DC-3. The airplane design had a nicely proportioned high aspect-ratio wing, with an additional feature that was very aesthetically pleasing: the wing was gracefully swept back. On a purely technical basis, this was a very unusual design feature. It came about when tests of a model of the original DC-1 design made at the California Institute of Technology wind tunnel showed the airplane to be unstable—the center of gravity of the airplane was behind the center of lift, a classic situation that causes uncontrolled pitching of the airplane. Douglas solved this problem by sweeping the wing backward, thus shifting the center of lift behind the center of gravity—a technical solution that was surely influenced by aesthetics, because other technical fixes were possible.

Great attention was paid to streamlining during the design of the airplane in order to reduce the aerodynamic drag as much as possible. Here the laws of nature were enhancing the aesthetics of the DC-3 (fig. 39). The net result of all these features was one of the most aesthetically

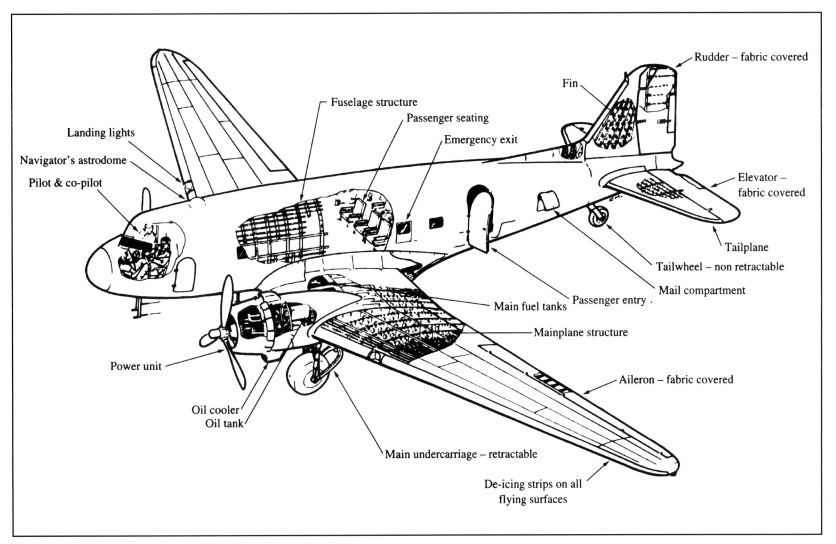

Rudder – fabric covered

Fin

Fuselage structure

Passenger seating

Emergency exit

Landing lights

Navigator's astrodome

Pilot & co-pilot

Elevator – fabric covered

Tailplane

Tailwheel – non retractable

Mail compartment

Main fuel tanks   Passenger entry

Mainplane structure

Power unit

Aileron – fabric covered

Oil cooler
Oil tank

Main undercarriage – retractable

De-icing strips on all
flying surfaces

39. View of the Douglas DC-3,
showing some of the internal
structures.

40. A Douglas DC-3 flies over the
Chicago skyline during regular
passenger service in the late 1930s.

pleasing airplanes ever built. Nothing in the technical literature from Douglas at that time hints that aesthetics was an overt player in the design of the airplane, but the subconscious feelings of the designers were certainly at work. The beautiful lines of the DC-3 also helped to make the public fall in love with the airplane, an unquantifiable competitive feature used to advantage by the airlines (fig. 40). The airline executives fell in love with the DC-3 for an additional reason: it outperformed all other commercial airliners of the time, and it made money for the airlines. The airplane was a classic example of form following function, another reason for its aesthetic beauty.

THE SR-71 BLACKBIRD

The historical evolution of the airplane has been dominated by the quest for more speed and higher altitude. The Lockheed SR-71 represents the epitome of this quest (fig. 43). In May 1965, this airplane set a new speed record of 2070 miles (3331 km) per hour, Mach 3.14, at an altitude of

42. A wind tunnel model of the
Lockheed SR-71 Blackbird,
1963–70. The SR-71 was developed
for the US Air Force by Clarence
"Kelly" Johnson's famous Lockheed
Skunk Works design bureau. First
flown on December 22, 1964, the
plane entered service with the
USAF Strategic Reconnaissance
Wing in 1966. Capable of flying at
greater than Mach 3 at altitudes
over 85,000 feet (26,000 meters),
the SR-71 served the US Air
Force in terms of reconnaissance
and NASA in terms of flight
research. Neither of them uses the
plane today. The last flight of a
Blackbird was in October 1999,
and the last aircraft was
decommissioned in 2002.

41. The sleek SR-71 in flight,
c. 1997. Designed by the Lockheed
Skunk Works in the 1960s for
reconnaissance purposes, it was
later used for research by NASA
throughout the 1990s.

43. The SR-71B, the only two-seat
pilot trainer model of a Blackbird,
seen in flight over California,
c. 1995.

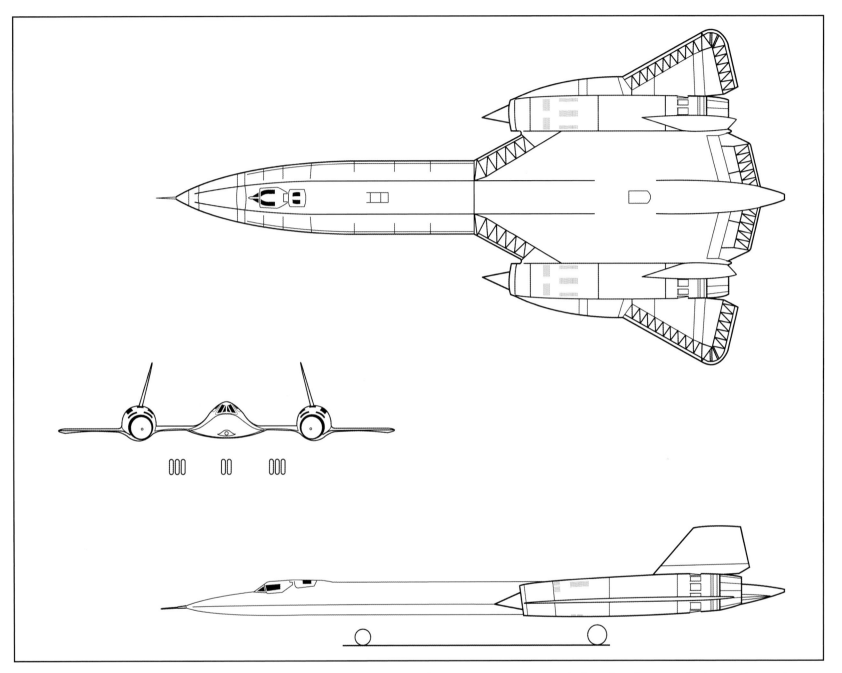

44. Schematic drawing of the
Lockheed SR-71 Blackbird aircraft.

80,258 feet (24,463 meters). Although the information is still classified, it is rumored that the Blackbird can exceed Mach 3.3. The Blackbird is unique—it is not the result of a normal evolution from a previously existing airplane. Its performance at such speeds and altitudes was a quantum leap from that of any past design. As the airplane's designer Kelly Johnson points out: "I believe I can truly say that everything on the aircraft from rivets and fluids, up through materials and powerplants, had to be invented from scratch."[11] The airplane was designed in the Lockheed Skunk Works, legendary for the innovative airplane designs that were produced there in the decades following World War II—

designs that always pushed technology to its limits. Kelly Johnson, arguably the most famous airplane designer since the Wright brothers, led the Skunk Works until his retirement in 1975.

The physics of air flow above Mach 1 (supersonic speeds) are completely different from that below Mach 1 (subsonic speeds). Nature simply plays with different rules at supersonic speeds, and for the SR-71 flying above Mach 3, these new rules entailed major design challenges. The shape of the Blackbird is unlike that of any previous airplane, yet it has a stark beauty to it (fig. 44). The long, slender fuselage and delta-shaped wing with thin airfoils reflect good supersonic

aerodynamics, designed to weaken the shockwaves that cause the large increase in drag—called wave drag—that dominates at supersonic speeds. In addition, at Mach 3 aerodynamic heating due to the friction of the high-speed air rubbing over the airplane's surface forced the designers to use titanium rather than the usual aluminum for the structure of the airplane. The engines were custom-designed by Pratt & Whitney specifically for the Mach 3 speeds of the Blackbird. Considering the cutting-edge technology that had to be used in order for the Blackbird to meet its performance specifications, aesthetics was probably the farthest thing from Kelly Johnson's mind (fig. 45).

One example of this aesthetic sensibility is the design of the Blackbird's chines, long narrow extensions of the wing that reach almost all the way to the nose of the fuselage. When moving

But Johnson had long years of design experience, including the beautiful P-38 Lightning from World War II, and the sleek, streamlined P-80 Shooting Star, the first mainline production jet airplane in the United States after the war. So Johnson had the subconscious "feeling" that made him a great designer. Although aesthetics was not in the design book for the SR-71, it was present, nevertheless, in the subconscious minds of the designers (fig. 46).

45. Many experiments at high altitudes and speeds in excess of Mach 3 were carried out using NASA's series of A-12, YF-12, and SR-71 aircraft over their various thirty-year carriers with NASA. The SR-71 was last used to obtain data on a proposed linear aerospike engine at altitudes and speeds in excess of Mach 1. Here a fit check is being performed between the SR-71 and the linear aerospike experiment (LASRE) in 1996. The last SR-71 flew in October 1999.

46. Model of the Lockheed YF-12 in the NASA Glenn Research Center's 10 × 10 foot (3 × 3 meter) supersonic wind tunnel during testing in 1980. Any proposed modifications, or the addition of external payloads or sensors, were wind tunnel tested before being flight-tested. The YF-12 is one of four vehicles in the SR-71 Blackbird family (the others are the A-12, the M-12, and the SR-71). Each is capable of sustained speed above Mach 3 and can attain altitudes in excess of 80,000 feet (24,380 meters).

48. Cross-section diagram comparing air crossflow over a round fuselage and over a fuselage with chines.

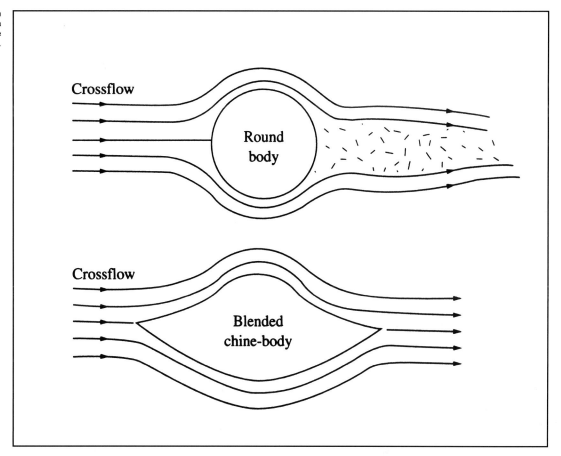

47. Student cockpit of the SR-71B, viewed from above; the SR-71B was purpose-built as a trainer. An instructor's cockpit is located above and behind the student. Blended nose chines, clearly seen in this view, provide lift for the forward fuselage and reduce the aircraft's radar cross-section. A pitot-static probe provides airspeed and pressure data to the cockpit instruments.

from subsonic to supersonic speeds an airplane experiences a dramatic shift of the center of lift (technically called the center of pressure). For a simple flat surface, the center of pressure shifts from a position near the plate's leading edge to halfway down it, and this shift is abrupt and dramatic as soon as the flow increases beyond Mach 1. In a more complex shape, such as an airplane, the sudden shift when moving from subsonic to supersonic speeds causes major changes in flight stability. On the SR-71, the technical solution was to add chines on both sides of the fuselage; this helped to minimize the center-of-pressure shift and made the stability problems manageable. When the SR-71 is viewed from above, the long chines add a degree of aesthetic beauty to the airplane; when viewed in cross-section, one appreciates the beauty of the differing curves on the upper and lower surface, and the way they join at a sharp, knife-like point on both sides of the fuselage (fig. 47). Moreover, the SR-71's chines, while aesthetically pleasing, also serve a technical purpose. As a result of the chines, the air flow over the side of the vehicle's fuselage (the crossflow) results in smoother streamlines when compared to the flow over a fuselage with a classic circular cross-section (fig. 48). So here is a

fine example of form following function: the laws of nature are put to optimum use, and result in an aesthetically beautiful configuration.

The distinguished aeronautical engineer Richard Passman (chief aerodynamicist for Bell Aircraft during the design of the Bell X-2 in the late 1940s and early 1950s, and later fully responsible for the design of the Bell X-16 concept) summed up the theme of this essay in a recent conversation with the author. He reflected:

There are few (if any) examples of compromise in aircraft design to satisfy aesthetics, but many times the resulting aircraft gives the designer a feeling of beauty accomplished. The final object reflects a balanced result of the desired objectives of lifting wings, control surfaces, and streamlined fuselage, which together accomplish the desired speed, altitude, and mission performance objectives.[12]

To this may be added the words of Charles C. Adams, Associate Professor of Engineering at Dordt College in the Netherlands. In his thoughtful article entitled "Technological Allusivity: Appreciating and Teaching the Role of Aesthetics in Engineering Design" (1995),[13] Adams states that

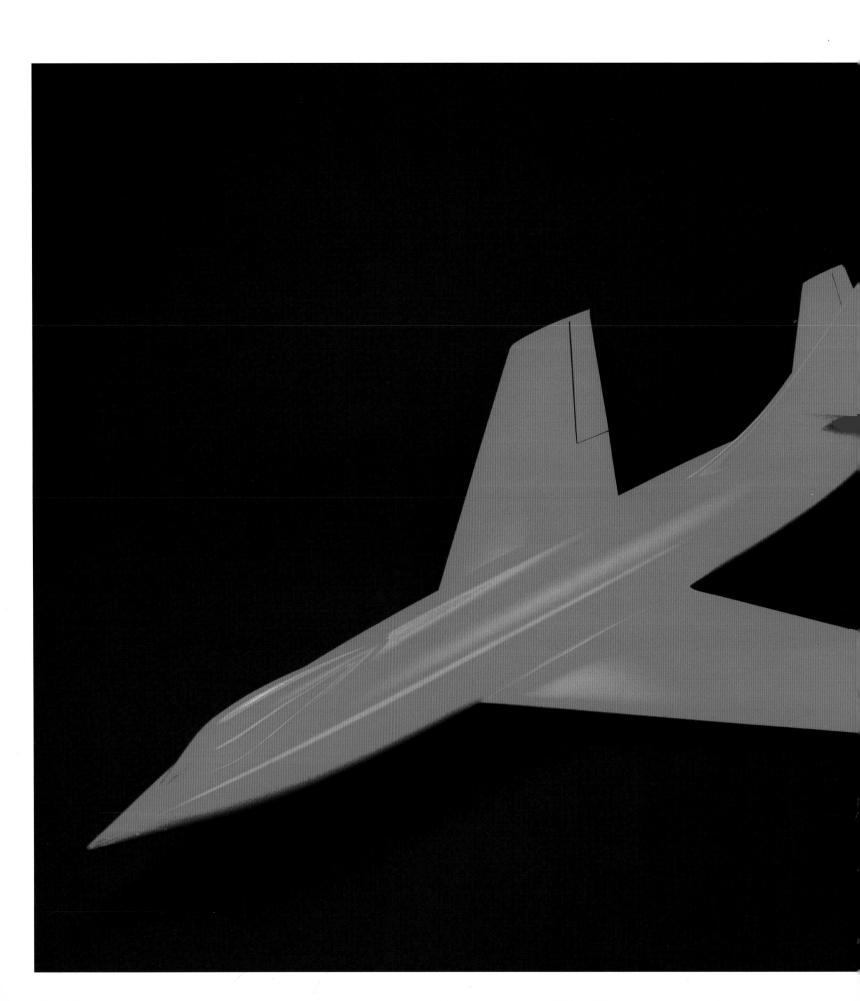

"good engineering design aesthetics implies (delightfully) harmonious interaction, at the human–technical interface, whereby the product dissolves into an extension of the user." This, of course, is one of the hallmarks of all good aerospace vehicles.

As we have seen, form follows function, with aesthetically pleasing designs resulting in successful vehicles. A vehicle's form must follow the laws of nature in order to succeed. Over time humans have found these forms to be aesthetically pleasing, but this not the case for vehicles that fly outside the realm of atmosphere. Compare the design of the lunar module, which was only used in the vacuum of space, with the design of a vehicle that would fly through the air (figs. 50, 51). We find the air vehicle's appearance satisfying, while the ungainly space vehicle only a mother would love. Will our idea of what is aesthetically pleasing change as we advance into space?

Notes

1. Harold Osborne (ed.), *The Oxford Companion to Art*, New York (Oxford University Press) 1970, p. 12.

2. Daniel P. Raymer, *Aircraft Design: A Conceptual Approach*, Restin, Va. (American Institute of Aeronautics and Astronautics) 1989, p. 1.

3. John D. Anderson, Jr., *Aircraft Performance and Design*, Boston (McGraw-Hill) 1999, p. 381.

4. Leonardo da Vinci's Notebooks, 1508–18, from a manuscript in the Royal Library at Windsor, England. See Edward MacCurdy (ed.), *The Notebooks of Leonardo da Vinci*, London (Jonathan Cape) 1938, vol. I, chapter 3.

5. Sir Thomas Browne, *Religio Medici* [1642], ed. Henry Gardner, Birmingham, Ala. (Classics of Medicine Library) 1981, part I, section 16.

6. Sir D'Arcy Wentworth Thompson, *On Growth and Form*, Cambridge, UK (Cambridge University Press) 1917.

7. C.P. Snow, *The Two Cultures: and a Second Look*, Cambridge, UK (Cambridge University Press) 1963.

8. Nevil Shute, *No Highway*, New York (Wiliam Morrow & Co.) 1947, reprinted London (Heinemann) 1963.

9. Richard Smith, "Better: The Quest for Excellence," in John T. Greenwood (ed.), *Milestones of Aviation*, New York (Hugh Lauter Levin Associates) 1989, p. 243.

10. Quoted in D.J. Ingells, *The Plane that Changed the World*, Fallbrook, Cal. (Aero Publishers) 1966.

11. Clarence L. Johnson, "Some Development Aspects of the YF-12A Interceptor Aircraft," American Institute of Aeronautics and Astronautics paper #69-757, New York, 1969.

12. Communication with the author, August 2002.

13. Proceedings of the American Society for Engineering Education/Institute of Electrical and Electronics Engineers Frontiers in Education Conference, November 2–4, 1995.

49. A Bell X-2 wind tunnel model, c. 1954. This gel-coated mahogany model was tested from Mach 1.0 to 1.5 in 1954–55. The actual Bell X-2 first flew under power in 1955, and was the first airplane to reach Mach 3, or three times the speed of sound—over 2,000 miles (3219 km) per hour. This record was achieved on September 27, 1956 at an altitude of 65,500 feet (19,964 meters); the pilot was Milburn "Mel" Apt. The X-2 also set an altitude record of 126,200 feet (38,466 meters) during a slightly earlier test flight piloted by Iven C. Kincheloe. Only two X-2's were built, and both were lost in accidents, reminding us of the tragic ends met by many of these machines and their test pilots as they strove to fly higher and faster.

50. Time-lapse exposure photograph taken during a night-time training session of the Lunar Excursion Simulator (LEMS) at NASA Langley, c. 1967. The LEMS was used by Apollo astronauts to train on earth for the landing of the lunar module on the moon. The lunar module did not look anything like an aircraft on earth since it was designed to fly in the vacuum of space and land in the 1/6 gravity of the moon. The LEMS was one of the many machines used to test out the aerospace systems used during the Apollo space program.

51. The lunar module after separation from the Apollo command module, in lunar orbit with astronauts Neil Armstrong and Buzz Aldrin aboard. The lunar module Eagle would ferry the astronauts safely to the first moon landing on July 20, 1969.

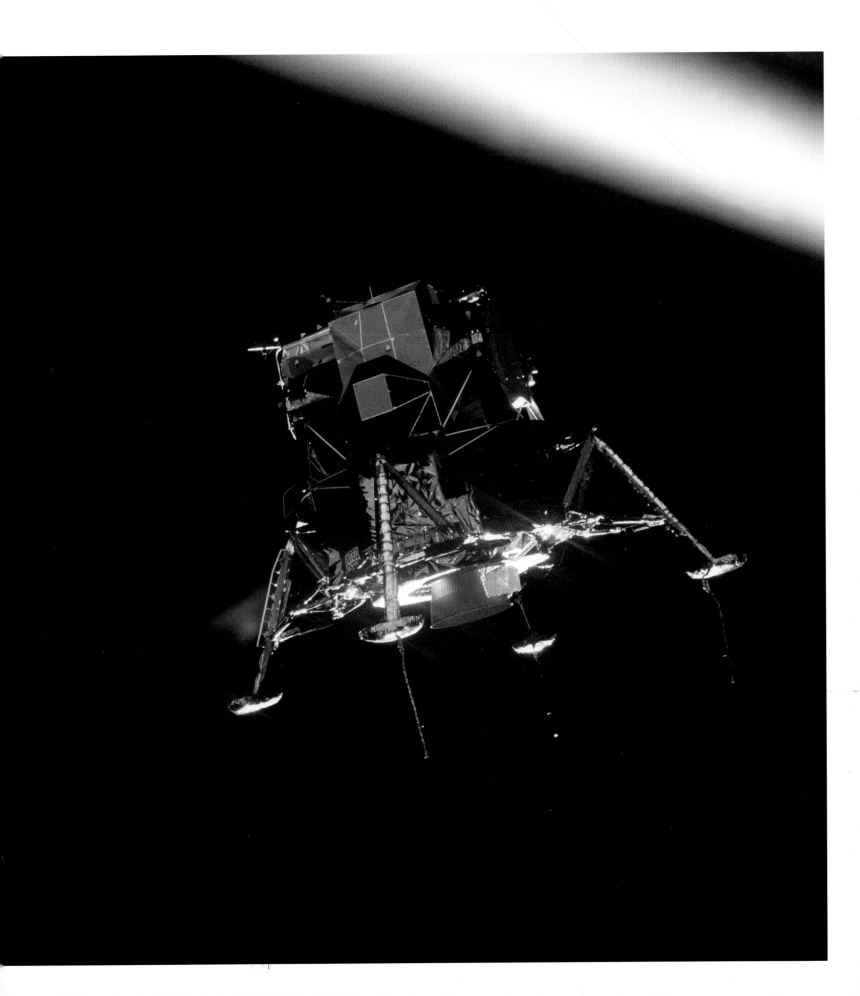

CHAPTER 04

James R. Hansen

BEAUTY IN THE BEAST:
FORM AND FUNCTION IN THE EARLY
DEVELOPMENT OF WIND TUNNELS

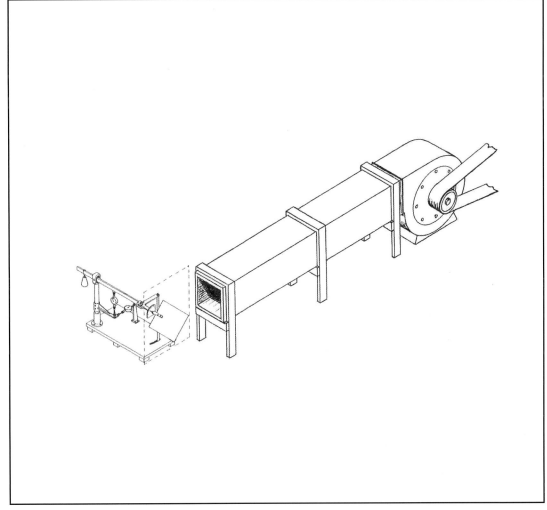

53. The world's first wind tunnel, built by Francis H. Wenham, a council member of the Aeronautical Society of Great Britain, in 1871. Data from Wenham's tunnel indicated much higher lift-to-drag ratios than those expected by Newtonian theory. His results suggested, among other things, that wings could support more substantial loads than previously expected. This added to the optimism that powered flight was attainable. This isometric drawing of Wenham's wind tunnel shows a square trunk attached to the end of a centrifugal fan. Part of the flat drive-belt is shown, as well as the balance apparatus (to the left) holding a flat plate (test object) at a 45-degree angle. (Drawing by J. Lawrence Lee.)

52. The passage of a body through the air can be visualized in many ways. One method, illustrated here, is the interferogram, which shows the changes in air density around an aluminum sphere as it is shot at supersonic speeds in a ballistic range.

LUKE: What a piece of junk.

HAN: … She may not look like much, but she's got it where it counts, kid.

Luke Skywalker and Han Solo, at Luke's first sight of the *Millennium Falcon*. From *Star Wars: Episode 4* (1977).

If an average person were to eyeball any of the world's first wind tunnels, the initial impression would be much like Luke Skywalker's harsh snap judgment of the *Millennium Falcon*. All of these pioneering aerodynamic test machines looked pretty ugly. Certainly, nothing about their outer appearance gave away the fact that here was a machine that would prove absolutely fundamental to the future progress of aeronautics.

British experimenter Francis H. Wenham built the world's first wind tunnel in 1871 (fig. 53). Aesthetically, the device was not much to look at, simply a 10 foot (3 meter) long wooden horizontal box measuring 18 inches (45.7 cm) square. Through it, a flow of air was driven at speeds up

to 40 miles (64 km) per hour by a small fan located at one end.[1] To the untrained eye, any number of everyday appliances from the Victorian age, from cotton gins to washing machines, look more complicated and interesting than Wenham's duct.

The world's next wind duct (the term "wind tunnel" was not coined until the early 1900s), built by another Englishman, Horatio Phillips, in 1884, looked only slightly less boxy. Its more streamlined appearance was due primarily to the presence of a cone-shaped steam ejector. Made from sheet metal with double-curved surfaces, the ejector connected to one end of a 6 foot (1.8 meter) long box in order to draw air into, through, and out of its test section.[2] No one mistook Phillips's creation for a sculpture by Rodin, who happened to be unveiling a new masterpiece in 1884.

The first tunnel in the United States came to life in 1896. So mundane was its appearance that an average person could not have distinguished a

54. Replica of the 1901 Wright wind tunnel. The Wrights called the machine a "trough." This apparatus was actually the second "trough" that they built. The first was only 18 inches (45.7 cm) long and built out of an old starch box. This version measured 16 inches (41 cm) square and was 6 feet (1.8 meters) long. Downstream from the fan there is a wire mesh and metal honeycomb straightener used to stabilize the flow.

cross-section drawing of it from a building's heating/cooling system. In fact, the engineering student at the Massachusetts Institute of Technology (MIT) who built the device, Albert J. Wells, used dampers in the basement ventilation ducts of a campus building to direct air to his tunnel. Once again, the machine—which Wells with appropriate modesty called "a pipe"—was hardly a thing of beauty.[3] Yet it remained in service until 1912 and helped a generation of MIT engineering students learn aerodynamics.

The same sets of adjectives—novel yet crude, ungainly but effective—describe the next two American wind tunnels. What distinguished the machine constructed in 1901 by Albert F. Zahm at The Catholic University of America in Washington, D.C., was its whopping 40 feet (12 meters) long overall size and 6 foot (1.8 meter) square test

section. Professor Zahm put his apparatus in a building that was to be devoted entirely to aerodynamic experiments. This did not make the homely looks of the blocky tunnel any more pleasing, but it did turn the overall facility into the world's first true wind tunnel laboratory.[4]

Several hundred miles to the west of the nation's capital, two brothers in Dayton, Ohio, also in 1901, constructed a smaller wind tunnel in the back room of their bicycle shop. Wilbur and Orville Wright called their device a "trough," and the baseness of the term captured the rudimentary character of their creation. Puzzled over the poor performance of their gliders at Kitty Hawk, and hoping to generate more accurate data about the lift of airfoils, the Wrights built the makeshift tunnel from an old starch box. A quickly assembled dwarf, it rested only 18 inches

None of the early wind tunnels, from Wenham's to the Wrights', was a beautiful piece of machinery in the sense of what people would normally associate with visual attractiveness or artistic elegance. But beautiful machines they nonetheless were, in that they unlocked many secrets of aerodynamics and flying. Their beauty depended not on cosmetics but on the way their function met demanding technological objectives. As American architect Louis H. Sullivan said in the early 1900s, "Form follows function."

OPEN-CIRCUIT DESIGNS

A significant factor in the form/function relationship of the earliest wind tunnels was their open circuitry; that is, they drew air into test passages directly from the atmosphere and released the air back into the atmosphere. None of them incorporated an extension for returning the air, turning the tunnel into a closed loop. The non-return-type atmospheric tunnel remained the norm into World War I. The tunnels built in and around Paris in the early 1900s by the great French engineer Gustave Eiffel still had open circuitry. This was true of Eiffel's 1.5 meter (45.6 cm) diameter tunnel at the Champs de Mars in Paris, completed in 1909, which sucked air through a test section at 20 meters per second—roughly 45 miles per hour. (Eiffel built the machine close to the already famous tower that bore his name so as to utilize electricity from the tower's generator to drive his tunnel's air blower.) It was also true of Eiffel's "open jet" tunnel at Auteuil, built in 1911–12. The improved design produced an airspeed of 32 meters per second—roughly 72 miles per hour (fig. 55).

To the untrained eye, Eiffel's tunnels looked only marginally more elegant than the tunnels that came before. But, like the *Millennium Falcon*, his machines had it "where it counts." To improve the flow of wind over the airfoils that he tested, Eiffel installed a convergent nozzle near the end of his tunnel. This stabilized the flow as air was sucked in. Near it he placed a honeycomb structure that also straightened the flow. He minimized the interference effects caused by the tunnel's solid walls by creating what today would be called a "semi-open" test section. Around a regular open test section he built a larger, hermetically sealed enclosure. The resulting "experimental chamber," as Eiffel called it, reduced wall interference, producing more accurate results.

55. Gustave Eiffel built his first wind tunnel in 1909 but conducted significantly more tests—over four thousand—in this second machine, built at his Auteuil laboratory outside Paris in 1911–12. It is not clear exactly how much his test results influenced the design of French aircraft into the World War I era, but there is no question that his data made a positive contribution.

(45.7 cm) long and was utilized only for one day, but that was long enough for the Wrights to verify that something was indeed wrong with the published aerodynamic coefficients (fig. 54). So informal were their tests that the brothers recorded data on scraps of wallpaper found lying on the floor of their workshop.

The Wrights went on to build a bigger and better "trough." This second apparatus, a wooden box 6 feet (1.8 meters) long and 16 inches (41 cm) square, was still quite simple, but by employing it in a series of systematic experiments over a period of two months—during which time they tested nearly two hundred airfoil models—the Wrights obtained information that proved critical to the remarkable success of their 1902 glider and its successor, the landmark powered airplane of 1903.[5]

Eiffel's tunnels still represented an immature stage of design. In them, nevertheless, the French engineer carried out the world's first wind tunnel tests of complete aircraft configurations, including wings, fuselage, tail, and landing gear, albeit in model form. The performance of his nation's aircraft, including the famed Nieuport fighter of World War I, benefited significantly from the systematic testing done in his tunnels.[6]

The first wind tunnel ever built and operated by the National Advisory Committee for Aeronautics (NACA), the US government's original civilian aeronautical research organization, established by Congress in 1915, copied what had become the "classic" open circuitry of the Eiffel tunnels. In fact, the NACA Wind Tunnel No. 1, built at Langley Memorial Aeronautical Laboratory in Tidewater Virginia in 1919, was a virtual copy of an Eiffel-type tunnel that the British had been operating at their National Physical Laboratory at Teddington, England, since 1916 (fig. 57).

The NACA's modest first machine—a non-return, atmospheric device with a 5 foot (1.5 meter) diameter test section—possessed a modicum of both inner and outer beauty. Its intricate honeycomb sections and mesh screens ensured good airflow quality, and its 200 horsepower electric motor enabled fairly precise control of tunnel air speed. A photograph taken in 1920 with two NACA mechanics posing near the entrance of the tunnel shows an attractive bell-shaped cone converging into the honeycomb structure (fig. 56). But as attractive as it was, the tunnel proved to be a bust. It allowed the NACA researchers to gain practice with the application of airfoil theory and the practical design of wind tunnel equipment, but enabled little else.[7]

CLOSED-CIRCUIT DESIGNS

What made the NACA Wind Tunnel No. 1 virtually obsolete even while it was being constructed was the earlier invention, in Germany, of the world's first closed-circuit tunnels. This new design brought a major leap forward in the effectiveness of a wind tunnel to simulate with scale models the actual conditions experienced in flight.

It came out of the University of Göttingen, where the great German physicist-engineer Ludwig Prandtl and his students in the early decades of the century made a number of key discoveries in fluid dynamics, including essential new ideas about airfoils, lift, drag, streamlining, skin friction, and boundary layer conditions. In 1908, Prandtl built the world's first continuous-circuit, return-flow machine at Göttingen. In it, the brilliant professor and his students tested airship, airplane, and airfoil models.

The closed-circuit tunnel offered distinct advantages over the open-circuit. It reduced power requirements through partial recovery of the kinetic energy of the air leaving the tunnel's diffuser. By incorporating improved screens and honeycombs, it produced and maintained a much more uniform airflow. It permitted pressurization and humidity control. To create a racetrack-like circuit for the airflow, Prandtl introduced efficient corner-turning vanes.

As effective as his 1908 machine proved to be, Prandtl quickly moved to improve his design. In 1916 he completed a second-generation closed-circuit tunnel that was so refined that it deserves the title of the first truly modern wind tunnel. A large settling chamber upstream of the test section further dampened airstream turbulence. A sharp contraction cone placed at the entrance to the test section greatly increased flow velocity. The turning vanes took the shape of more effective airfoils. Altogether, the new machine produced faster, smoother, drier, and more reliable airflow than any earlier tunnel. The design was so efficient that a 300 horsepower motor, rotating a four-bladed fan, produced a wind velocity of 170 feet (52 meters) per second or roughly 116 miles (187 km) per hour—higher than any previous speed in a wind tunnel.[8]

Keeping close tabs on German developments, the British built their first closed-circuit device, with a 4 × 4 foot (1.2 × 1.2 meter) test section, at the National Physical Laboratory in 1910. Curiously, the next British closed-circuit machine did not come until 1929.[9] The first American machine of this type came in 1913. Designed by David W. Taylor and Albert Zahm at the Washington Navy Yard, the machine was unusual for a closed circuit in having a test section that was larger in its dimensions (8 × 8 feet / 2.4 × 2.4 meters) than those of the rest of the circuitry. It was also unprecedented in being the first of many tunnels to be built and paid for by the US government.[10]

56. Two wind tunnel researchers
pose near the entrance end of
Langley's 5 foot (1.5 meter)
Atmospheric Wind Tunnel (AWT).
Air was pulled into the test section
through a honeycomb arrangement
meant to smoothen the flow.

57. The NACA's first wind
tunnel, Tunnel #1 or the 5 Foot
Atmospheric Wind Tunnel (AWT),
built at the NACA Langley Memorial
Aeronautical Laboratory (now the
NASA Langley Research Center). It
was designed by an in-house team,
led by Edward P. Warner and
Frederick H. Norton, and became
operational on June 11, 1920. In
1930 it was dismantled following
a series of minor revisions.

THE VARIABLE DENSITY TUNNEL

The next decisive step in the evolution of the wind tunnel was an American development, but one inspired by a German insight. In early 1920 Dr. Max M. Munk, a Prandtl protégé, wrote from Göttingen to the National Advisory Committee for Aeronautics in Washington, D.C., stating that he had found an answer to the problem of "scale effects." Everyone working in aeronautical research had known for a long time that a scale model of an airplane placed into a wind tunnel could not generate results precisely comparable to the aerodynamic forces actually experienced by a full-scale machine in flight. Test results could be "scaled up" using empirical coefficients, but the air itself, with its properties of density and temperature, could not be "scaled down" with any accuracy to model size. To gain maximum value from wind tunnel testing, an answer to the scaling problem had to be found.

Dr. Munk offered to tell the NACA his answer if the US government would give him a job and bring him to the United States. Desperate for a solution and impressed with his credentials, the NACA employed Munk as a technical consultant.[11]

Munk advised the NACA to build a tunnel *inside* a pressure vessel so that tests could be run under high pressure, thereby increasing the density of the air.[12] Under Munk's supervision, the NACA built exactly such a facility at its Langley laboratory in Virginia in 1921–22. Known as the Variable Density Tunnel (VDT), it was only the NACA's second tunnel but the results it produced turned out to be superior to any obtained previously in a wind tunnel (figs. 58, 59).[13]

The VDT looked monstrous. From the outside it appeared only as a large cylindrical tank, with small windows cut into it to enable viewing of the test section during operation and a hatch at one end providing access to the interior. The real beauty, as with nearly all of these machines, lay inside: a 5 foot (1.5 meter) diameter tunnel in which air flowed through a central test section and then past a fan that blew the air back around via an annular return passage. With the tank pressurized to 300 pounds (136 kg) per square inch, or the equivalent of twenty atmospheres, the tunnel produced Reynolds numbers equivalent to full-scale flight.

58–59. The NACA first built its reputation as an outstanding aeronautical research institution on the strength of its Variable Density Tunnel (VDT). The tank for the VDT arrived at Langley by rail from its manufacturer, the Newport News (Va.) Shipbuilding & Dry Dock Company, in February 1922. It was an 85 ton (86 tonne) pressure shell with walls made from steel plate that was lapped and riveted according to a practice standard in steam-boiler construction.

NACA
18683

60. Thanks to an energetic technical apprenticeship system, boys in the neighborhoods surrounding NACA laboratories learned to turn their boyhood passion for building airplane models into a profession. Here young men make models to be used in Langley's Free-Flight Tunnel in November 1939.

Higher Reynolds numbers were the key to truly effective wind tunnel testing. In the 1880s British engineer Osborne Reynolds established experimentally that transition from smooth fluid flow to turbulent fluid flow always occurred when certain variables exceeded a critical value defined by a specific flow parameter. The result came to be known as the "Reynolds number." Because it was dimensionless, the Reynolds number could be used to compare fluid-flow forces around similarly shaped, but differently sized, objects. One could achieve what came to be called "dynamical similarity" by varying different parameters, such as

decreasing the velocity or increasing the density, to produce the same Reynolds numbers for different tests.[14]

With its tank pumped up to twenty atmospheres, the ungainly-looking VDT produced extraordinarily high Reynolds numbers for its day. It produced data very close to that generated by a full-size aircraft in flight. This led the NACA into a comprehensive program of unprecedented airfoil research. Starting in the mid-1920s, the laboratory began cranking out report after report on airfoil performance. Published catalogs of systematic test data enabled the fledgling aircraft

61. Model makers install flaps and wiring on the wing of a 1/12th scale model of the Martin JRM-1 flying boat in the Langley Dynamic Model Shop in 1944.

industry to find just the right shape that would work best for the wings and propellers of its new aircraft designs.

The models tested in the VDT were built at 1/20th scale in order to match up with the twenty atmospheres of pressure. Over the years at the NACA laboratory, model making became an art form, one involving stringent geometrical and material requirements. Many wind tunnels, including the VDT, needed models made of metal, preferably duralumin, to withstand high dynamic pressures and powerful vibrations. Those that did not require metal had their own demands. If not built properly, wooden models could break up, especially at high angles of attack, sending splinters and other debris flying through a tunnel. Early tests in the VDT confirmed that models

had better replicate the exact geometry of the full-scale model or test data would be highly unreliable.

So important was model making to successful aeronautical research that the NACA very early on instituted an in-house apprentice program to train its own staff of expert craftsmen. In their precisely sculptured aerodynamic contours and handsomely varnished wood or polished metal, wind tunnel models possessed a physical beauty that often contrasted sharply with the ugly mechanical features of the wind tunnel itself (fig. 61).

Adding to the artistic side of a wind tunnel was its array of lift-and-drag balances, gauges, model support systems, and other delicate instruments required to stabilize the test model and record all

62. NACA mechanics install the new NACA cowling for testing on an airplane during 1928.

the performance data. Like other reasonably sophisticated mechanical, electrical, and electro-mechanical equipment, wind tunnel instruments, when well designed and nicely integrated into the overall machine, struck engineers and technicians as beautiful, if not elegant. What gave this technology its aesthetics were the ingenious ways in which meticulously designed and crafted devices did what they had to do.[15]

TOWARD THE MODERN AGE

From the late 1920s onward, wind tunnels grew significantly larger and more powerful. This changed the drama of the machine. Up until 1927, no tunnel anywhere possessed a test section larger than 10 feet (3 meters) in diameter. But in 1927 the NACA built a 20 foot diameter

machine, the Propeller Research Tunnel (PRT), which deserves to be called the world's first large wind tunnel.

No one called the PRT pretty. Although the big machine had dual return circuits, it simply used the exterior walls of the building that contained it as part of its circuit wall. What impressed people about the PRT was its size and power. An actual airplane minus its wing tips could be mounted in its test section. Two 1000 horsepower engines salvaged from navy submarines drove the wind flow to a speed of 160 feet (48.7 meters) per second, or 110 miles (177 km) per hour. One of the first research programs carried out in this tunnel led to the design of a low-drag engine cowling, for which the NACA won its first Collier Trophy in 1929 (figs. 62, 63).[16]

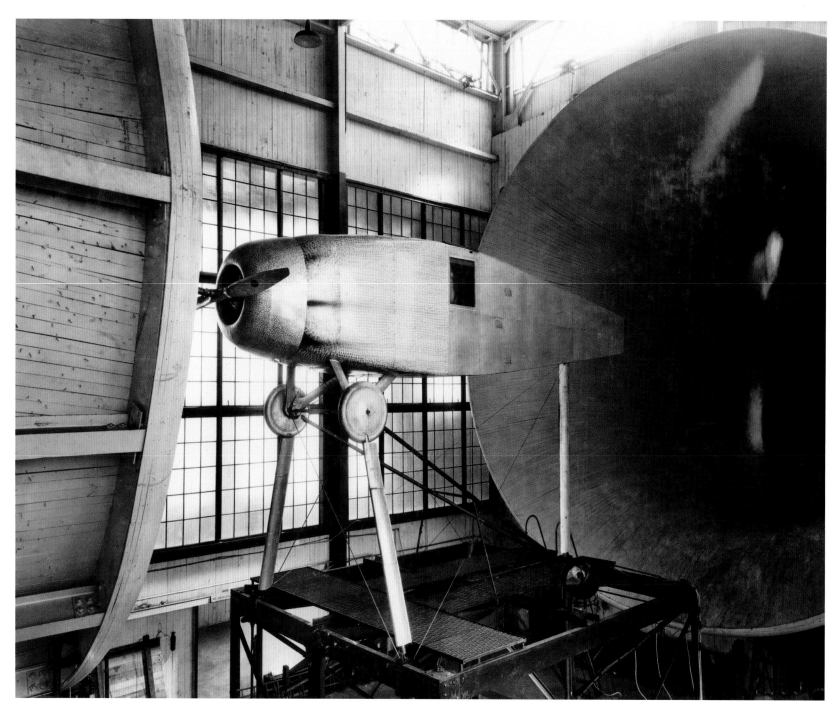

The PRT inspired the NACA to build an even larger tunnel, one with a 30 × 60 foot (9.1 × 18.3 meter) test section (fig. 64). This facility opened at Langley in 1931 as the Full-Scale Tunnel (FST). It marked the first time that a complete airplane with its full wingspan could be tested in a wind tunnel. A closed-circuit machine, the FST proved gigantic in every way. Two monstrous 35 foot (10.7 meter) diameter propellers in the dual return ducts, each driven by a 4000 horsepower electric motor, circulated almost 160 tons (163 tonnes) of air through the 838 foot (255 meter) long circuit and produced wind velocities of nearly 120 miles

(193 km) per hour. Because scale factors were minimized, these speeds enabled measurements that could be confidently extrapolated to cover the aircraft's entire speed range. Inside the FST— which is still operating today, some seventy years later—researchers tested not just an immense array of different aircraft but also airships, helicopters, submarines, spacecraft, and automobiles (figs. 65, 66).[17]

Large tunnels like the FST came to loom over most aerodynamic laboratories. At its new Ames laboratory in California, the NACA in the early

63. By late 1928, tests of cowling no. 10 in Langley's Propeller Research Tunnel (PRT), illustrated here, showed a dramatic reduction in drag. Since the earliest days of flight drag has presented a problem for aircraft, and much of Langley's early research was focused upon reducing aircraft drag. One method was to place a cowling or covering over the engine cylinder heads, much like the hood over the engine of a car. The NACA was to receive the prestigious Collier Trophy for the development of this cowling.

64. A Vought O3U-1 Corsair is tested in the Langley 30 × 60 foot Full-Scale Wind Tunnel in 1931. This was the first aircraft to be tested in the full-scale tunnel, and is seen here during preliminary testing before the balance mechanism was enclosed. NACA engineers compared the lift and drag characteristics obtained in the wind tunnel with those acquired from flight tests. The Full-Scale Tunnel had been designed by a team led by Smith J. DeFrance, Abraham Silverstein, and Clinton H. Dearborn. It became operational in 1931, underwent major reconstruction in 1977, and later transferred to Old Dominion University, Norfolk, V.I.

1940s built a mammoth tunnel that was 40 × 80 feet (12 × 24 meters) at the test section (fig. 67). To outsiders, a complex of tunnels like that at Langley and Ames may have looked like a swarm of huge, wormlike creatures, washed ashore perhaps after a battle of primordial monsters in a nearby body of water. But the tunnels were no less fascinating to those whose gaze was less imaginative. The exteriors of big wind tunnels may have looked only like large warehouse buildings with jointed appendages and rounded corners, but they were in fact complicated mechanized marvels, national

resources, great and powerful monuments to the modern age. Whenever the NACA hosted a major conference at Langley or Ames, as it did often, a group picture was always taken in the cavernous test section of the 30 × 60 or 40 × 80 foot tunnels. The photo would be taken there not just because it was the only place large enough to hold a big group but also because no other facility provided such an impressive visual backdrop and evoked so dramatically the power of technology.[18] It was like having one's picture taken in front of Hoover Dam or the Golden Gate Bridge.

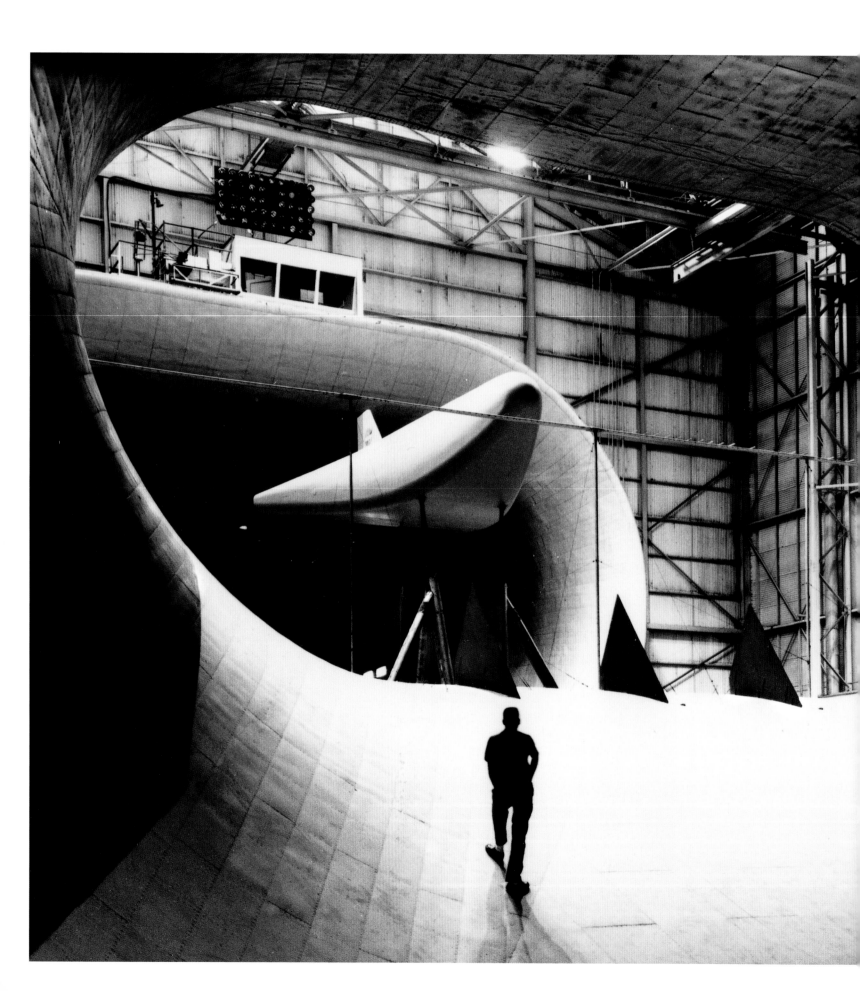

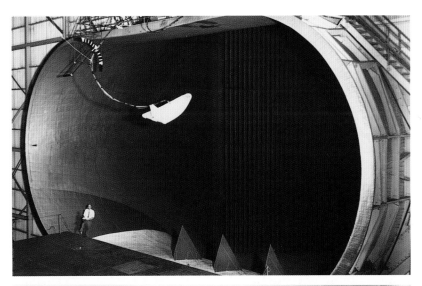

66. (top) The free flight testing of an HL-10 lifting body model in the FST. The model is remotely "flown" in the wind tunnel to determine the low speed dynamic stability and control characteristics of the vehicle. The cable connecting the vehicle to the tunnel carries power to the model's control surfaces.

67. (above) A full-scale replica of the 1903 Wright Flyer in the NASA Ames Research Center's 40 × 80 (12 × 24 meter) Wind Tunnel during 1999. The original Ames subsonic tunnel was contructed in 1944 as a closed-circuit tunnel, and an 80 × 120 foot (24 × 37 meter) leg was added in 1982 (see fig. 77) to test full-scale aircraft at low speeds. The picture above juxtaposes the advances made in one hundred years in flight. The Wright Flyer, the first true aircraft, is tested in a machine—the wind tunnel—that has been built to test modern aircraft designs. The tunnel is longer than the length of Wrights' first flight, and the aircraft is mounted on a pylon higher than it ever flew.

68. (following pages) A model of the experimental Convair XF2Y Sea Dart undergoing testing in the tow tank at NASA Langley in the mid-1950s. The XF2Y Sea Dart was designed to be the first supersonic seaplane, and first flew in December 1956. Langley's tow tank, enclosed in the "mile-long" building, allowed models to be pulled along the surface of the water using an overhead track. Tow testing was used to determine the interactions between the water and the proposed aircraft, spacecraft, or boat. Tests were done on seaplanes for take-offs and landings, on planes ditching at sea, and for the recovery of spacecraft.

65. A subscale model of the HL-10 lift body aircraft is tested in the Langley Full-Scale Tunnel (FST). The lift body aircraft has no wings, the lift needed for flight being generated by the shape of the body.

70. (top right) The pathway of the airflow for the 16 Foot Transonic Wind Tunnel at NASA Langley is shown in this artist's sketch of the facility. This closed-loop or continuous flow wind tunnel recycles the air, which flows over the model mounted in the test section. The name for the wind tunnel is derived from the tunnel test section size, 16 feet (4.9 meters) on a side, and the speed range of which the tunnel is capable. Transonic is at or near the speed of sound, 760 miles (232 km) per hour at sea level.

71. (right) The Rogallo Wing, a flexible glider developed as a means to recover spacecraft or boosters, is tested in the 7 × 10 foot (2.1 × 3 meter) wind tunnel at NASA Langley. Invented by Francis Rogallo, these flexible wings were in fact never to be used for spacecraft recovery, but they did contribute significantly to the birth of the sport of hang gliding. The wing's deployment is being studied in this test.

69. This scale model used in sonic boom research studies at NASA Langley was no bigger than the human eye. Only 1 inch (25 mm) in length, the model was used to study the pressure field associated with supersonic aircraft.

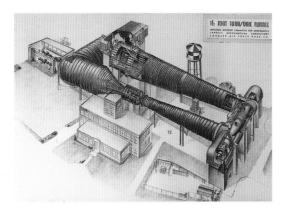

72. An advanced turboprop engine model with propeller swirl recovery additions is shown in the NASA Glenn Research Center's 8 × 6 foot (2.4 × 1.8 meter) Supersonic Wind Tunnel. This scale model was tested to determine if vanes aft of the advanced turboprop propeller blades could be used in place of a set of counter rotating blades. The 8 × 6 Supersonic Wind Tunnel, designed in-house by John Disher, Abraham Silverstein, and Marquis D. Wyatt and constructed by the American Bridge Company, became operational in 1949.

Many of our world's greatest technological artifacts are themselves fine art. The engineers who designed these structures applied scientific principles, but they also used practical skills, cleverness in contrivance, and an innate sense of aesthetics. Although science and mathematics also play important roles in technology, artistic vision fixes many of the outlines and fills in many of the design details. But not all artistry involves pretty pictures. Much artistry is found in ingenious solutions to practical problems, and in finding highly effective forms to carry out functions.[19]

The design of the wind tunnel falls into that latter category. In the seventy-plus years since the NACA created the Full-Scale Tunnel, dozens of other wind tunnels have been built, in countries all over the world. (The construction of no fewer than 135 tunnels has been documented to the year 1941; a good estimate might be that three to four times that many have been built since.) These machines simulate flying conditions at subsonic, transonic, supersonic, and hypersonic speeds (fig. 70). They allow researchers to explore flying conditions in the air, in space, and in the

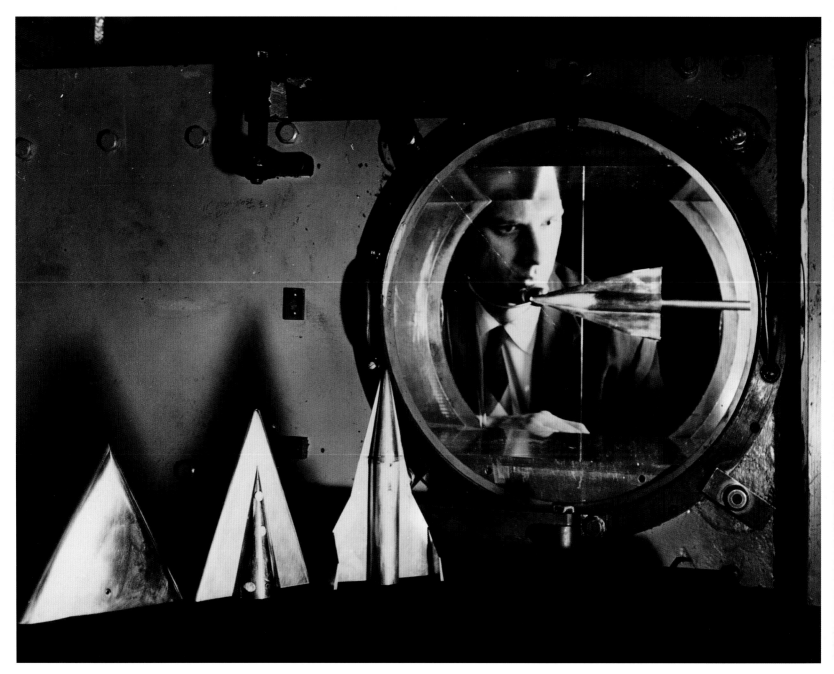

trans-atmospheric conditions in between. The tunnels come in all sorts of sizes, shapes, and dimensions (figs. 69, 71–73). Today, these devices are more powerful, versatile, reliable, specialized, and precise than ever before. Sophisticated computer programs have been simulating aerodynamic flow conditions for more than twenty years now, but they have not replaced the wind tunnel. Nature and the real world of flying have proven too complicated for such modeling (fig. 74).[20]

The wind tunnel thus remains a critical tool. But as a machine it has "morphed" into a post-modern form—one that benefits from extensive computerization, laser-based instrumentation, cryogenic operation, magnetic model suspension, and self-streamlining walls, in which the very shape of a tunnel's walls changes by design. In the twenty-first century, wind tunnels will undoubtedly prove to be no more attractive outwardly than the pioneering forms of the late nineteenth and early twentieth centuries (figs. 75, 76). Their form will continue to evolve as engineers look to build more effective machines by which to understand the many still mysterious functions of machines flying through the air (fig. 78).

73. Scientists and engineers studied the aerodynamic characteristics of proposed hypersonic rocket-boost glide vehicles in the Langley 11-Inch Hypersonic Wind Tunnel during the late 1950s. Models could be tested at Mach 10 using air as the operating fluid for the tunnel, and Mach 18 using helium.

74. A technician inspects the guide vanes of the NASA Langley Research Center's 16 Foot Transonic Wind Tunnel. The facility, which was designed by David J. Beirmann and Lindsey I. Turner, opened in 1941. The guide vanes at each corner of the closed-circuit tunnel smooth the air and guide it around the 90-degree turns. If it were not for these guide vanes, the air would pool in the corners—as a stream does around the corner of a bank. The tunnel has been renovated and improved many times, most recently in 1990.

75. The 16 Foot Wind Tunnel at NASA Langley. Constructed in 1940 as the high-speed wind tunnel, it was later modified into a transonic wind tunnel.

76. In NACA Langley's trailblazing 8 Foot High-Speed Tunnel, which became operational in 1936, researchers for the first time investigated the severely adverse stability and control problems encountered by airplanes in high-speed dives. The concrete walls of the igloo-like structure seen to the left of the photograph needed to be one foot thick in order to withstand the powerful, inwardly directed pressure generated by the tunnel. Conceived by Eastman N. Jacobs and designed by Russell G. Robinson and Manley J. Hood, the tunnel was reconfigured to simulate transonic speeds in the late 1940s and deactivated in 1956.

77. NASA Ames Research Center's
80 × 120 foot (24 × 37 meter)
Full-Scale Wind Tunnel, one of the
largest in the world, is used to test
full-scale aircraft at low speeds.
Shown here is the exit.

I would like to thank two individuals in particular for the contributions they have made over the years to my understanding of wind tunnels and their history: Donald D. Baals, retired NASA engineer and co-author of *Wind Tunnels of NASA* (NASA SP-440, 1981), and James Lawrence Lee, author of "Into the Wind: A History of the American Wind Tunnel" (PhD diss., Auburn University, 2001). Dr. Lee completed his dissertation in the history of technology under my direction after many years as a working mechanical engineer.

78. The air flow around a model of the F-16XL is seen using smoke and a laser light sheet during a wind tunnel test in NASA Langley's Basic Aerodynamic Research Tunnel (BART), 1993. Lasers are used to visualize the flow of air over the model, thus aiding the development of vehicle configuration. These techniques are used to analyse the flow field around a vehicle before proceeding to flight-test.

79. (following pages) An engineer inspects the 16-foot-long fan blades of the 16 Foot Transonic Wind Tunnel at NASA Langley in the early 1950s. The blades, driven by a 16,000-horsepower electric motor, propelled a strong current of air around the closed-loop tunnel to speeds of just over and just below the speed of sound.

Notes

1. On Wenham's wind tunnel, see N.H. Randers-Pherson, "Pioneer Wind Tunnels," *Smithsonian Miscellaneous Collections 93*, January 19, 1935, pp. 1–2. On Wenham's contributions to aerodynamics, see John D. Anderson, *A History of Aerodynamics and Its Impact on Flying Machines*, Cambridge, UK (Cambridge University Press) 1997, pp. 116–17 and 119–26.

2. Randers-Pherson, *op. cit.* (n. 1), p. 3. See also Anderson, *op. cit.* (n. 1), pp. 238–39, and James Lawrence Lee, "Into the Wind: A History of the American Wind Tunnel," PhD diss., Auburn University, 2001, pp. 21–24.

3. Albert J. Wells, "An Investigation of Wind Pressure upon Surfaces," BS diss., Massachusetts Institute of Technology, 1896, pp. 4–23.

4. See A.F. Zahm, "New Methods of Experimentation in Aerodynamics," paper presented at the meeting of the American Association for the Advancement of Science, Pittsburgh, June 20, 1902, in *Aeronautical Papers of Albert F. Zahm, Ph.D., 1885–1945*, South Bend, Ind. (University of Notre Dame) 1950.

5. See Peter Jakab, *Visions of a Flying Machine: The Wright Brothers and the Process of Invention*, Washington, D.C. (Smithsonian Institution Press) 1990, pp. 122–24, and Tom D. Crouch, *The Bishop's Boys: A Life of Wilbur and Orville Wright*, New York and London (W.W. Norton) 1989, p. 222.

6. For an English translation of Eiffel's own description of his wind tunnels, see G. Eiffel, *The Resistance of Air and Aviation: Experiments Conducted at the Champs-de-Mars Laboratory*, trans. Jerome C. Hunsaker, London (Constable & Co.) and Boston (Houghton Mifflin & Co.) 1913. For an assessment of Eiffel's tunnels and their contributions to aerodynamics, see Anderson, *op. cit.* (n. 1), pp. 267–82.

7. On the history of the NACA and its first wind tunnel, see Alex Roland, *Model Research: The National Advisory Committee for Aeronautics*, Washington, D.C. (NASA SP-4103) 1985, especially vol. I, p. 83 and vol. II, p. 514; and James R. Hansen, *Engineer in Charge: A History of the Langley Aeronautical Laboratory*, Washington, D.C. (NASA SP-4305) 1987, pp. 69–72 and 442.

8. For the history of the aerodynamics research organized under Ludwig Prandtl's leadership at the University of Göttingen, see Paul A. Hanle, *Bringing Aerodynamics to America*, Cambridge, Mass. (Massachusetts Institute of Technology Press) 1982. Those who read German should consult Julius C. Rotta, *Die Aerodynamische Versuchsanstalt in Göttingen*, Göttingen (Vanderhoeck & Ruprecht) 1990, especially pp. 45–47. For a detailed analysis of Prandtl's contribution to aerodynamics, see Anderson, *op. cit.* (n. 1), pp. 251–60.

9. See "Specifications for Major Wind Tunnels, 1871–1941," in Lee, *op. cit.* (n. 2), pp. 191–92.

10. J. Norman Fresh, "The Aerodynamics Laboratory—The First 50 Years," *Aero Report 1070*, Washington, D.C. (Department of the Navy) 1964, pp. 7–14.

11. Hansen, *op. cit.* (n. 7), pp. 72–78 and 84–95; Roland, *op. cit.* (n. 7), pp. 87–98.

12. Max M. Munk, "On a New Type of Wind Tunnel," *NACA Technical Note 60*, Washington, D.C. (NACA) 1921. Also Eastman N. Jacobs and Ira H. Abbott, "The N.A.C.A. Variable Density Wind Tunnel," *NACA Technical Report 416*, Washington, D.C. (NACA) 1932.

13. Hansen, *op. cit.* (n. 7), pp. 78–84; Anderson, *op. cit.* (n. 1), pp. 301–04 and 342–48.

14. See Osborne Reynolds, "An Experimental Investigation of the Circumstances which Determine Whether the Motion of Water in Parallel Channels Shall be Direct or Sinuous, and the Law of Resistance in Parallel Channels," *Proceedings of the Royal Society*, London 1883, pp. 84–89. On Reynolds's contribution to fluid mechanics, particularly the concept of the "Reynolds number," see Anderson, *op. cit.* (n. 1), pp. 109–14.

15. For a summary of early wind tunnel balances, recording instruments, and associated equipment, see George W. Gray, *Frontiers of Flight: The Story of NACA Research*, New York (Alfred A. Knopf) 1948, chapter 3: "Tunnels and Other Tools," pp. 34–62.

16. Fred E. Weick and Donald H. Wood, "The Twenty-Foot Propeller Research Tunnel of the National Advisory Committee for Aeronautics," *NACA Technical Report 300*, Washington, D.C. (NACA) 1928. On the design, construction, and early operation of the PRT, see Fred E. Weick and James R. Hansen, *From the Ground Up: The Autobiography of an Aeronautical Engineer*, Washington, D.C. (Smithsonian Institution Press) 1988, pp. 49–59. For a historical analysis of the NACA cowling program, see Hansen, "Engineering Science and the Development of the NACA Cowling," in Pamela E. Mack (ed.), *From Engineering Science to Big Science: The NACA and NASA Collier Trophy Research Project Winners*, Washington, D.C. (NASA SP-4219) 1998, pp. 1–28. The latter is an expanded version of the chapter on the cowling in Hansen, *op. cit.* (n. 7).

17. Smith J. DeFrance, "The NACA Full-Scale Wind Tunnel," *NACA Technical Report 459*, Washington, D.C., 1933. See also Donald D. Baals and William R. Corliss, *Wind Tunnels of NASA*, Washington, D.C. (NASA SP-440) 1981, pp. 22–23, and Hansen, *op. cit.* (n. 7), pp. 101–05, 194–202, and 447–49.

18. On the history of the NACA annual aircraft manufacturers' conferences, see Hansen, *op. cit.* (n. 7), pp. 148–58.

19. Eugene S. Ferguson provides a wonderfully insightful analysis of the role of artistry in technological design in his book *Engineering and the Mind's Eye*, Cambridge, Mass. (Massachusetts Institute of Technology Press) 1992. Ferguson expressed his essential ideas on this subject fifteen years earlier in his article "The Mind's Eye: Nonverbal Thought in Technology," *Science* 197, August 26, 1977, pp. 827–36. On the role of the artistic in technological creativity, I also recommend Robert Pirsig, *Zen and the Art of Motorcycle Maintenance*, New York (William Morrow & Co.) 1974; Arnold Pacey, *The Maze of Ingenuity: Ideas and Idealism in the Development of Technology*, Cambridge, Mass. (Massachusetts Institute of Technology Press) 1985; and three short books (published in New York by Farrar, Strauss, and Giroux) by John McPhee: *The Deltoid Pumpkin Seed* (1973), *The Survival of the Bark Canoe* (1975), and *The Curve of Binding Energy: A Journey into the Awesome and Alarming World of Theodore B. Taylor* (1976).

20. The best single survey of the wind tunnel up to modern times is Baals and Corliss, *op. cit.* (n. 17). Unfortunately, as the title implies, the book mainly looks at NASA tunnels. Another useful survey of tunnel development, though only through the 1960s, is Kenneth L. Goin, "The History, Evolution, and Use of Wind Tunnels," *AIAA Student Journal*, vol. 9, February 1971. On the persistence of the wind tunnel in the face of Computational Fluid Dynamics (CFD), see Anderson, *op. cit.* (n. 1), pp. 441–45, as well as Baals and Corliss, *op. cit.* (n. 17), pp. 136–41.

CHAPTER 4, APPENDIX 1

AN INDISPENSABLE TOOL

To engineer is to foresee the different ways in which a technology can fail. Successful engineering only happens when myriad opportunities for failure are effectively avoided.

No tool has proven more indispensable to avoiding failure in modern aeronautical engineering than the wind tunnel. Whether it involves testing an aircraft's lift, drag, stability, controllability, or some other aspect of its performance, wind tunnel research simulates how an aircraft or aircraft component will fly, and it does so without risking the real thing or costing excessive amounts of money. Piloted free-flight tests of the full-scale aircraft will eventually need to be made, but only after the effectiveness of the flight vehicle's detailed design has been scrutinized in a program of wind tunnel experiments. No manufacturer in the world, today or fifty years ago, would commit the many millions of dollars it takes to develop a new airplane without relying on a large body of systematic tunnel data.

One story from the history of flight is enough to underscore the wind tunnel's importance. Late on the evening of September 29, 1959, Braniff Flight 542, a Lockheed turboprop known as Electra, crashed near Buffalo, Texas. The plane went down after its left wing failed, killing all thirty-four persons on board. Six months later, on March 17, 1960, another Electra, operated by Northwest Orient, mysteriously disintegrated over Tell City, Indiana. All sixty-three people on the airliner died (fig. 80).

When investigators found no signs of metal fatigue in the remains of either Electra, aeronautical engineers began to suspect that the accidents were due to wing flutter. This was a highly dangerous phenomenon, not well understood at the time, in which an aircraft or one of its components—in this case, its wings—absorbed energy from the airstream in which it was moving and converted it into a harmonic oscillation embedded within its own structure. In certain rare conditions, this vibration could grow so strong that the structure collapsed from it. Aerodynamicists and structures experts had been exploring flutter for decades when the Electra crashes occurred. But their only insights were highly theoretical. Predicting when flutter might happen was an arcane science based on complex mathematical theories that only a few geniuses could formulate and the average engineer could not even understand. What was necessary to resolve the uncertainty was some hard data about the various modes of structural vibration and the ways in which an airplane could succumb to flutter.

Fortunately, in early 1960, a new wind tunnel was just going into service at NASA's Langley Research Center, one specially built for flutter investigations. Conceived eight years earlier by former NACA researchers that were now part of NASA, the Transonic Dynamics Tunnel (TDT) offered the world of flight its first true aeroelastic-test environment.

NASA researchers faced an urgent situation. Some 130 Electras were still flying, though at reduced cruising speeds. The aircraft industry needed an answer, and fast, or more tragedies could occur. The public needed to regain its confidence not just in that particular airplane but also in the safety of flying generally. Everyone looked to NASA's new wind tunnel for the answer.

What theory could not analyse or predict, the TDT pinpointed in only a few weeks. The gyroscopic forces associated with the Electra's engine–propeller combination led to a wobbling motion that had a frequency that exactly matched the natural flutter frequency of the wings. In a manner similar to what had happened in the collapse of the Tacoma Narrows Bridge in 1940, oscillations from the harmonic resonance could become so powerful that the wings fell off. The answer found in wind tunnel testing was to strengthen the engine mounts. Apparently the downed Electras had engine mounts weakened by storms or hard landings. By strengthening the mounts, the fatal resonance could be avoided.[1]

Having solved the mystery, the job of the Transonic Dynamic Tunnel became to make sure that the Electra's story was never repeated. In the years that followed, virtually all new military aircraft and civilian airliners went through systematic flutter investigations in the TDT. NASA researchers found the causes of the tail flutter of the Lockheed C-141 military transport, of the McDonnell Douglas F-15 fighter's horizontal tail flutter, and of the General Dynamics F-16's wing flutter. Without the wind tunnel, many more aircraft might have met with catastrophe.

Not all wind tunnel testing solves problems so directly or dramatically, of course. Researchers use the devices to test the design of aircraft more generally, to generate basic aerodynamic information, and to develop and refine aerodynamic theories. As the veteran aerospace engineer, Dr. John Anderson, writes, "Aerospace engineering in general, and aerodynamics in particular, is an empirically based discipline. Discovery and development by experimental means have been its lifeblood." The "workhorse" of empirical research in aeronautics has been the wind tunnel, "so much so that today most aerospace industrial, government, and university laboratories have a complete spectrum of wind tunnels ranging from low subsonic to hypersonic speeds."[2] Not even the revolutionary development of computer hardware and software capabilities can replace the tunnel (figs. 81, 82).

Without question, the wind tunnel will continue to play a crucial role in aerospace research and development for many decades into the future.

Notes

1. See Donald D. Baals and William R. Corliss, *Wind Tunnels of NASA*, Washington, D.C. (NASA SP-440) 1981, pp. 80–81.
2. John D. Anderson, *Introduction to Flight*, 3rd edn., New York (McGraw-Hill) 1989, p. 159.

80. Lockheed Electra Prop Jet. An early victim of wing flutter, the Lockheed Electra was the only large turboprop airliner to be developed in the US. Although highly efficient, the Electra did not compete successfully with the contemporaneous Boeing 707 jet airliner.

81. (right) Extensive wind tunnel testing is undertaken for every launch vehicle, despite the fact that they pass through the air extremely quickly. Here the Space Shuttle launch vehicle is tested in the ascent configuration, with metal bodies used to simulate the plumes from the main engines and the solid rocket boosters.

82. (far right) A generic airplane model, called the Pathfinder, is tested in the National Transonic Facility to determine the effects of model size or scale effects on the model's aerodynamic characteristics. This model is used in conjunction with an identical but larger companion to determine what effect size has on the aerodynamic data generated using the tunnel.

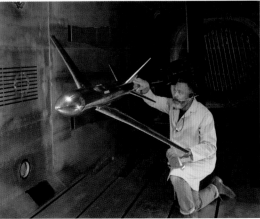

CHAPTER 4, APPENDIX 2

THE POST-MODERN TUNNEL

The wind tunnel has come a long way since the little homemade machine built by the Wright brothers in a back room of their bicycle shop in 1901. It has evolved from a simple little device made for less than $100 out of an old cornstarch box into a large and incredibly complex machine, costing nearly $100 million, that requires a bank of sophisticated high-speed computers to run it. If Ludwig Prandtl's closed-circuit tunnel of 1916 deserves to be called the first truly modern wind tunnel, the title of first "post-modern" tunnel must go to NASA's National Transonic Facility (NTF), formally dedicated at Langley Research Center in 1983. So complex was the design and construction of this revolutionary facility that it took more than thirty years for NASA to move from its basic conceptualization to the time the machine started producing its first wave of new aerodynamic information (figs. 83, 84).

The NTF was a conventional wind tunnel only in respect to its basic circuitry: a single-return, fan-driven tunnel, with a 2.5 meter (8.2 foot) slotted-wall test section. In virtually all other respects, it broke new ground. Its 120,000 horsepower electric drive included a two-speed gear that turned a fan incorporating controllable pitch. Pressures inside its shell could be made to vary from a near vacuum to nine atmospheres, and test gas temperatures could be adjusted from −300° to 175° F (−184° to 79° C). The machine could operate with air as the test medium or it could operate cryogenically (*i.e.*, with fluids or conditions at low temperatures, usually at or below −150° C, or 123 K), with 1200 pounds of liquid nitrogen injected into its circuitry every second. Four fine-mesh screens located in the settling chamber assured high-quality flow, and thousands of square feet of strategically located sound-absorbing panels sought to make it the quietest transonic facility operating anywhere in the world.[1]

Only someone deeply familiar with the state of the art in wind tunnel technology at the end of the twentieth century could truly appreciate the NTF's beauty. The tunnel benefited from the most advanced computerization. Four separate banks of high-speed computers controlled and monitored the tunnel's every operation as well as that of the test models. Even during short runs of the tunnel, which were typical in the cryogenic mode, the NTF's computers acquired and displayed data from nearly four hundred channels at rates as high as 50,000 points per second.

Every brand new piece of technology has some kinks to work out, and the NTF faced its share of early problems. But by the early 1990s, with new fan blades and a new 101-megawatt drive system installed, the NTF proved capable of generating the extremely high Reynolds numbers necessary for accurate aerodynamic simulation (as high as 143 million). In the "Big Cold One," as some insiders called the tunnel because of its cryogenics, researchers addressed the effects of Reynolds numbers on a wide range of critical phenomena, including aeroelasticity, buffet onset, aileron effectiveness, nacelle pylon interference, and much more. Results from the studies indicated that some phenomena associated with transonic aerodynamics behaved as expected but that others definitely did not. Nearly everything about Reynolds number effects proved to be dependent on the specifics of an aircraft's configuration, whether one was dealing with subsonic transports, supersonic transport concepts, or high performance military aircraft. Models tested in the NTF included a 65-degree delta wing, the Boeing 767, the X-29, numerous subsonic transports, different configurations of a High-Speed Civil Transport (HCST), and many more. In carrying out a very wide range of tests, the NASA facility fulfilled its mission as a national resource for industry, academe, and the Department of Defense (fig. 83).[2]

Other countries followed the US's lead and also built advanced wind tunnels designed to learn more about the vexations of flight at transonic speeds. In the early 1980s, researchers at the University of Göttingen, in the tradition of Prandtl, built a small cryogenic tunnel—a special type known as a "Ludwieg tube"—that managed to simulate transonic flows and achieve Reynolds numbers of nearly 60 million. Along with this facility, known as the "KRG," the Germans also built other innovative new machines. The British opened several new transonic facilities of their own, as did the French. The Dutch and the Swedes followed suit. A union of countries cooperated to build a "European Transonic Wind Tunnel" near Amsterdam. By 1993, twenty-two different cryogenic wind tunnels were operating around the world, including one in the People's Republic of China, five in Japan, two in Russia, and one in the Ukraine.[3]

This international flourishing of new facilities devoted to transonic research, many of them cryogenic, clearly demonstrated the continuing importance of the wind tunnel as an essential instrument for learning about flight. What sort of machine will come after the "post-modern" wind tunnel is anybody's guess. But the sky is the limit when it comes to technological inventiveness and humankind's ambition to know how to make things work.

Notes

1. See Donald D. Baals and William R. Corliss, *Wind Tunnels of NASA*, Washington, D.C. (NASA SP-440) 1981, pp. 133–36.

2. Information about the recent operations can be accessed at http://wte.nasa.gov/facility/ntf. This website is regularly updated.

3. The Von Kármán Institute at Saint Genèse, Belgium, sponsored an international colloquium on cryogenic tunnels in June 1989. See AGARD Report 812, "Special Course on Advances in Cryogenic Wind Tunnel Technology," Brussels (AGARD) 1989. (AGARD is the North Atlantic Treaty Organization's [NATO's] Advisory Group for Aeronautical Research and Development.)

83. Not all models tested in wind tunnels are of aircraft. Here a model of a Los Angeles Class Attack Submarine, mounted upside-down, is tested in the National Transonic Facility (NTF) at NASA Langley Research Center. Designed by NASA in conjunction with Sverdrup, and built mainly by Chicago Bridge & Iron Co., the NFT became operational in December 1983.

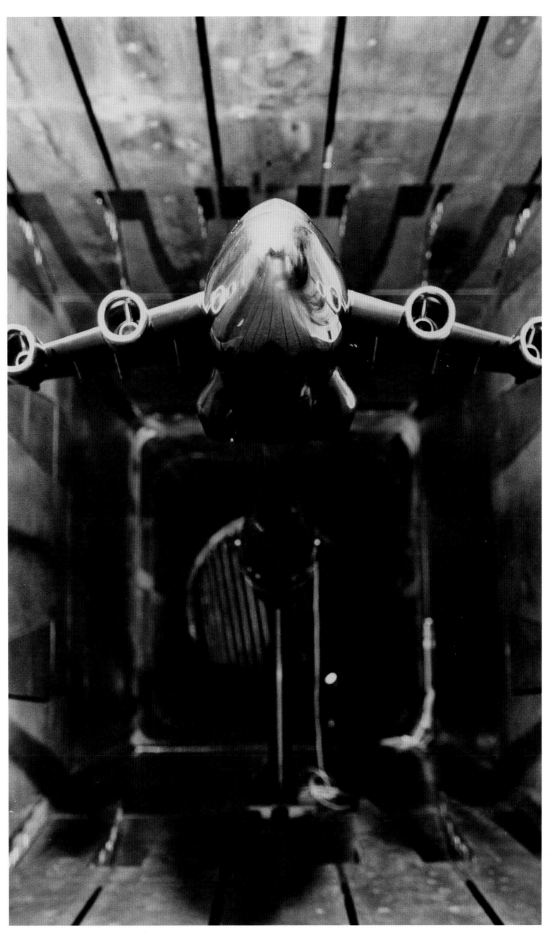

84. A C-17 High-Wing Military Transport Model is tested in the National Transonic Facility, NASA Langley Research Center.

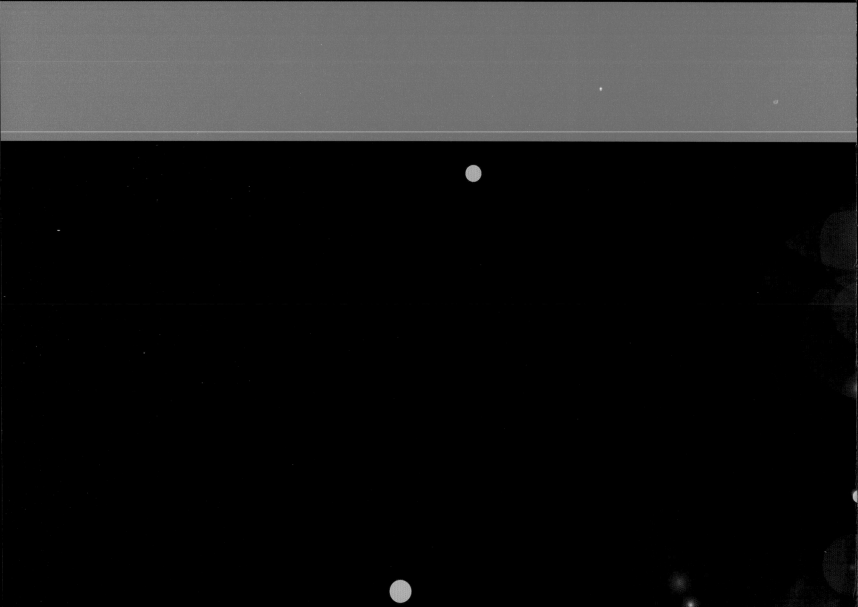

Flight research—the art of flying actual vehicles in the atmosphere in order to collect data about their behavior—has played a historic and decisive role in the design of aircraft. Naturally, wind tunnel experiments, computational fluid dynamics, and mathematical analyses all informed the judgments of the individuals who conceived of new aircraft. But flight research has offered moments of realization found in no other method. Engineer Dale Reed and research pilot Milt Thompson experienced one such epiphany on March 1, 1963, at the National Aeronautics and Space Administration's Dryden Flight Research Center in Edwards, California.[1] On that date, Thompson sat in the cockpit of a small, simple, gumdrop-shaped aircraft known as the M2-F1, lashed by a long towline to a late-model Pontiac Catalina. As the Pontiac raced across Rogers Dry Lake, it eventually gained enough speed to make the M2-F1 airborne. Thompson braced himself for the world's first flight in a vehicle of its kind, called a lifting body because of its high lift-to-drag ratio. Reed later recounted what he saw:

Slowly Milt brought the nose of the little lifting body up until the M2-F1 got light on its wheels. Then, something totally unexpected happened. The M2-F1 began bouncing back and forth from right to left. Milt stopped the bounce by lowering the nose, putting weight back on the wheels. Several times he again brought the nose up until the M2-F1 was light on its wheels, and each time the vehicle reacted the same way, Milt ending the bounce by lowering the nose as he had the first time. Later, in our little debriefing room, Milt said that he felt that if he lifted the M2-F1 off its wheels, it would have flipped upside down in a roll.[2]

Of course, the M2-F1 did not take wing without considerable preliminary research. Models of it had been tested in one of the wind tunnels at the NASA Ames Research Center in California, advanced mathematical analyses had predicted the aircraft's stability and control, and a simulator animated by an analog computer approximated its flying qualities.[3] But none of these disciplines uncovered the phenomenon that imperiled Milt Thompson's life (fig. 85).

The engineers and scientists who design aircraft—typically for large manufacturers—weigh not only the hard data of these fields but also a dizzying number of nontechnical variables. These other factors include military and commercial requirements, the prevailing design philosophy of

85. NASA research pilot Milton O. Thompson stands in front of the first of the lifting bodies, the M2-F1, on Rogers Dry Lakebed at Edwards Air Force Base. Fashioned simply from wood and steel tubing, the M2-F1 enabled the lifting body concept to be tested at low cost.

87. One of the pioneers in the field of compressibility, Dr. Hugh L. Dryden also made fundamental contributions to boundary layer research. During the final years of his career he became one of the leading architects of the American space program.

86. Aerial view of the NACA's flight research facility, not long after its founding on Muroc Army Air Base in California's Mojave Desert. The NACA hangar is in the foreground on the right, and the administrative offices in the long building to the right of the hangar. Parked in front of the hangar is much of the NACA's flying inventory.

their own firms, and the costs associated with the materials and production techniques. In addition to these vagaries, the designers' own talents and preferences influence the ultimate appearance of the machines rolling off the assembly lines.

In the final analysis, however, technical considerations commonly take precedence in the delicate calculus of trade-off and compromises. In the pursuit of hard data on which to base design, flight research has been an essential ingredient, in part because its practices are time-honored, beginning as much as a century before the Wright brothers. Indeed, the most prominent of the nineteenth-century aeronauts—Sir George Cayley, Otto Lilienthal, and Octave Chanute, among others—all employed its principles. So did Orville and Wilbur Wright. Flight research then, as now, enabled researchers to verify the behavior of aircraft as they flew in the air and to derive concrete data based on the motions of the vehicles aloft. Contrary to common belief, those involved in flight research did not merely *test* unproven designs. Hugh L. Dryden, director of the National Advisory Committee for Aeronautics (the NACA) from 1947 to 1958 and the first deputy administrator of NASA from 1958 to 1965, said that flight research existed to "separate the real from the imagined problems and to make known the overlooked and the unexpected problems" (fig. 87). Indeed, flight research may yield a variety of data, depending on the objectives of the particular project. In addition to separating predictions from reality and discovering the unanticipated, its practitioners sought to unravel the physiological effects of the cockpit on human beings; to perfect the art of flight safety in order to protect the lives of pilots, crews, and passengers; and to hasten cooperation and technology transfer among government and industrial entities. Most importantly, those involved in flight research attempted to comprehend the phenomena underlying the behavior of vehicles in flight, and then to report their findings in the technical literature.[4] Therefore, as aeronautical engineers fashioned their designs, they routinely augmented in-house data with the broad and systematic knowledge published in the reports and memoranda of the NACA and NASA (fig. 86).

Among the varied ingredients employed by the aircraft industry, the data derived from flight research represented a unique body of aeronautical experience. No designer could afford to ignore information derived from actual flight in the atmosphere, although flight research has not *necessarily* reshaped design. Sometimes, it simply confirmed the soundness of a proposed concept or feature. But oftentimes, it occasioned pivotal revisions that prompted wholesale changes in a generation of aircraft that followed. In some instances, the transformations yielded rapid redesigns; more often than not, it took time for the new ideas to be adopted.

Moreover, many of the new designs resulting from flight research could not be detected by sight at all. Between the mid-1930s and the mid-1940s, the NACA Langley Memorial Aeronautical Laboratory in Hampton, Virginia, flew a series of experiments crucial to aircraft handling qualities (fig. 88). The need for these tests became evident during the early 1930s. Until then, aeronautical engineers usually concentrated on the presence of stability in their designs. But as aircraft—especially airliners, transports, and bombers—became bigger and more powerful, pilots reported increasing difficulty maintaining control as they wrestled for hours with sluggish, relatively unresponsive sticks and wheels in the cockpit.

Frequently, aviators who flew the heavy planes arrived at their destinations exhausted mentally as well as physically, a fact exacerbated by the capacity of the new aircraft to fly longer ranges and to carry more passengers. Clearly, the safety of the flying public demanded that the balance between stable flight and responsive handling be redressed in favor of more nimble control. As a consequence, the Langley engineers conducted research on such borrowed planes as a Boeing B-17 bomber from the Army Air Corps and a Douglas DC-3 airliner from United Airlines, as well as a host of other vehicles. Gradually, these researchers evolved a standard "vocabulary" of handling qualities, based mostly on pilot assessments of the "feel" necessary to achieve a sense of ease of control. Published in the NACA technical reports, the research influenced cockpits the world over.[5] As a result, the next generation of big planes flew with far greater reference to human physiology, although the improvements themselves left little external sign of the internal adaptations (fig. 89).

88. Aerial photograph of the Langley Memorial Aeronautical Laboratory, located in Hampton, Virginia, taken in 1953. Seminal flying-qualities research occurred in the skies over this facility during the 1930s and 1940s, ultimately leading to a revolution in cockpit design. The NACA had been a tenant on the Army's Langley Field since 1917, but from 1946 high-speed flight research transferred to Muroc Army Air Base in Southern California.

VISIBLE AND FAST RESULTS

Many of the breakthroughs yielded by flight research involved obvious redesigns of the contours of aircraft; some have been more subtle. Essentially, readjustments have occurred at one of two intervals: either during the flight research program itself, or in subsequent months or years. A classic case of rapid change occurred on the famous Bell X-1 aircraft. In flying the world's first aircraft to surpass Mach 1, the US Air Force (USAF) and the NACA actually owned two separate versions with a distinct difference. The Air Force's model (the X-1 number 1) featured unusually thin wings (8% airfoil thickness to chord length). In contrast, the NACA's X-1 number 2 came equipped with significantly thicker wings (10% thickness to chord) (fig. 90). After several years of highly instrumented flying, NACA researchers concluded that the slender airfoil caused less turbulence both in the transonic and supersonic regions. As a consequence, the NACA leaders in Washington, D.C.—in collaboration with engineers at the NACA's newly established High-Speed Flight Station (the forerunner of today's NASA Dryden Flight Research Center)—decided to test the thin-wing concept with a new round of flight research (fig. 91). To do so, they

90. A DC-3 airliner on the ramp at the Langley Laboratory, 1937. Loaned by United Airlines to the NACA, the DC-3—along with many other aircraft—took part in lengthy and extensive flying-qualities research during the 1930s and 1940s.

89. The NACA's Bell X-1, shown with some of its relaxed ground crew. Known as the X-1 number 2 to distinguish it from the US Air Force's Mach 1-breaking X-1 number 1, the NACA airplane featured thicker wings than did the USAF version. It was later transformed into the X-1E, with wings that were thinner than those of the USAF plane, and its flight program tested the result.

National Advisory Committee for Aeronautics

NACA

High-Speed Flight Station

proposed a transformation of the X-1 number 2, rendering it a virtually new vehicle. Designated the X-1E, it looked considerably different to its predecessor. In addition to its extraordinarily slender wings (measuring 4% thickness to chord), it incorporated a cockpit canopy and an ejection seat. (Invisible to casual observers were its highly modified XLR-11 engines, which operated on a low-pressure system fed by a turbine pump.) Equipped with modifications borne out during the X-I's flight research program, the X-1E flew twenty-six times between 1955 and 1958 in a much broader performance profile than the earlier plane. It eventually achieved Mach 2.2 and offered important data about the aerodynamics likely to be encountered at hypersonic speeds.[6] Its thin wings became common on subsequent supersonic aircraft (fig. 92).

Another pivotal and visible redesign resulting from flight research involved the transfiguration of the M2-F1 (Modification 2-Flight Version 1) lifting body from a humble glider to the heavyweight M2-F2, and ultimately the M2-F3. The lifting body concept—first introduced by NASA Ames aerodynamicist H. Julian Allen in 1950—theorized that a rounded, blunt body shape re-entering the atmosphere from space would be resistant to incineration because its aerodynamics would enable the intense heating to be dissipated around the descending vehicle, rather than on its surface. Seven years later, a team of Ames researchers decided that a wingless vehicle shaped like a cone halved at its lengthwise axis offered the best chance to behave like Allen's generalized blunt body under actual flight conditions. During the early 1960s, a young Dryden engineer named Dale Reed embraced the idea of translating the Ames concept into a flying

91. The staff of the NACA's High-Speed Flight Station pose in front of the newly opened main building in 1954. Specializing in flight research, in 1976 it became the Hugh L. Dryden Flight Research Center.

92. Under the immense belly of
its B-29 mothership, the X-1E
aircraft—a more powerful, thinner
winged version of the X-1
number 2—receives attention
from a group of technicians.

machine (fig. 93). A radio-controlled-aircraft enthusiast, Reed fabricated and personally flew several models. He then persuaded the center's legendary research pilot Milt Thompson to lend his prestige to the project, an important factor in persuading Dryden's director Paul Bikle to back Reed's initiative with some discretionary funding. The money bought a simple little bathtub-like machine weighing only 1000 pounds (453.6 kg), fashioned from tubular steel and a wooden shell. It flew for the first time in March 1963, with Milt Thompson in the cockpit. Its propulsion system consisted of a souped-up Pontiac Catalina, pulling the aircraft by cable as it raced over the hard-packed surface of Rogers Dry Lake. The M2-F1's initial flight gave Thompson a truly frightening ride (see page 109). After a series of remedial tests in the Full-Scale Wind Tunnel at Ames, the M2-F1 returned to the skies with somewhat better handling qualities, and

93. Lifting Body Project engineer Dale Reed holds a model of his M2-F1, with the actual flying machine behind him. Reed is largely responsible for designing this first lifting body, and for launching a twelve-year flight research investigation of the lifting body family of aircraft.

94. The M2-F1 in one of its many tow flights behind the C-47 transport. The M2-F1 exhibited serious roll instability in flight.

95. (top) A worthy successor to the M2-F1, the heavyweight M2-F3 is seen above the surface of Rogers Dry Lake on Edwards Air Force Base. The M2-F1 and the M2-F2 taught engineers a great deal about lifting body design, resulting in the relatively stable and easier-to-fly vehicle shown here.

96. (above) The ultimate lifting body, the USAF-NASA X-24B, in flight over Edwards Air Force Base. Totally redesigned from earlier lifting body planforms, it proved to be a superb aircraft to fly.

eventually flew through the air while being towed by a C-47 aircraft (fig. 94). While it exhibited an unmistakable tendency to roll, by the time the M2-FI retired from the skies in August 1966 its many glide flights had taught Dale Reed and the other engineers how a true re-entry vehicle might be designed.

Indeed, the concept showed enough promise for NASA headquarters to fund a powered, fully flyable aircraft to succeed the M2-F1. The Northrop M2-F2, built to Reed's precise specifications, differed immensely from its predecessor. Fabricated of metal instead of wood and propelled by the XLR-11 engine (the powerplant flown on the early X-15 flights), it weighed four times as much as the M2-F1 and included sophisticated flight-control systems. Unfortunately, it not only looked like the M2-F1 but also flew like it. In May 1967, the aircraft crashed, gravely injuring research pilot Bruce Peterson, owing to the same problem as that fought by Thompson in the M2-F1: uncontrolled lateral oscillations. But the original lifting body design did not remain grounded. Virtually rebuilt, the M2-F3, which would fly in the transonic and supersonic ranges of speed, now featured two critical design changes, one visible, the other not. The unseen change was to add a computerized stability-augmentation system in order to tame some of the wingless aircraft's less mannerly handling qualities. In addition, borrowing from another lifting body known as the HL-10, the project engineers added a middle vertical fin between the two outboard ones for improved roll control (fig. 95).

This important and highly apparent modification, in tandem with the computer-aided controls, tamed the beast. In earlier transonic flights, and then in its first supersonic flight in August 1971, the M2-F3 showed itself capable of satisfactory flying qualities.[7] Thus, the M2 design adapted itself to the lessons of flight research and eventually demonstrated the worthiness of glide return from space. Perhaps more important, data from the M2s informed the designs of two highly successful lifting bodies that followed: the USAF's bulbous X-24A, and the long, sharp-nosed X-24B (fig. 96).

A MORE DIFFUSE INFLUENCE

Despite the obvious and rapid evolution of the X-1E from the X-1, and the M2-F3 from the

M2-F1, more often than not the influence of flight research on aircraft design has been somewhat diffuse. Moreover, the physical contours of the new aeronautical discoveries commonly emerged over a period of time, rather than right away.

A long and consequential research program pursued by the NACA during the 1920s illustrates the life-cycle of this more subtle influence of flight research on design. As the speed and agility of aircraft—particularly military aircraft—increased throughout the 1920s, combat and test pilots witnessed a disturbing trend: increasing numbers of their colleagues plunged to their deaths in airplanes that disintegrated in flight. No one could be sure why these events occurred, much less what could be done about it. At this point, Captain James H. ("Jimmy") Doolittle, then a young Army Air Service pilot, took it upon himself to launch an investigation of this frightening phenomenon (fig. 98).

Doolittle undertook this task with impressive qualifications. Educated at Berkeley as an undergraduate, in 1924 he searched for a thesis topic to satisfy the requirements of a Masters degree in aeronautical engineering from the Massachusetts Institute of Technology.[8] Doolittle found his subject at McCook Field in Dayton, Ohio. There he put an Air Service Boeing PW-9— a solid, durable plane—through a series of punishing maneuvers. Acting as his own flight research pilot and chief engineer, he hoped to arrive at some preliminary conclusions about the effect of air pressure on aircraft structures

98. Famed aeronaut James H. Doolittle photographed around the time of his grueling and dangerous pressure distribution flights during the 1920s.

97. Three of the NACA leaders during the 1930s: left to right, Henry Reed, engineer in charge of the Langley Laboratory; Vannevar Bush, chairman of the NACA; and George Lewis, the NACA's director of research. The affable and able Lewis, a strong proponent of flight research, served as NACA director from 1919 to 1947.

subjected to intense g (gravity) forces. The initial evidence persuaded him that pilots took the greatest risk to the integrity of their vehicles (and their lives) when they pulled up abruptly from steep dives.[9] This discovery proved to be life-saving knowledge for military aviators.

Yet, the military services reacted to Doolittle's findings with alarm. If his report proved to be accurate, the entire fleet of US combat planes might be subject to structural collapse during standard combat conditions. Indeed, at this point in the history of aeronautical research, no one could say which aircraft might suffer this fate and exactly what changes in design and fabrication might be required to avert more crashes in the future. Hence, in September 1924 (just six months after Doolittle flew his hair-raising sorties at McCook), the Air Service turned to the NACA for assistance. The positive response by NACA director George W. Lewis—a resourceful leader and an engineer of some experience—launched

his organization on one of its longer and more productive flight research projects (fig. 97).

Known as the pressure-distribution investigations, these tests lasted from 1925 until 1932 at the Langley Laboratory. Like Doolittle, the NACA fielded a PW-9 for the flights, but this one had been strengthened structurally by Boeing to NACA specifications. The PW-9 was one of the few aircraft actually purchased—rather than borrowed—by the NACA since its founding in 1915, and its presence signaled the importance that George Lewis and the Langley engineers ascribed to this project. Moreover, in this instance they outdid their own tradition of careful instrumentation of research vehicles. All of the PW-9's main control surfaces—wings, horizontal stabilizer, elevator, vertical fin, and rudder—had been implanted with manometers and accelerometers capable of moment-to-moment, simultaneous stress analyses (fig. 99).

99. Seen in front of a NACA hangar, this PW-9 aircraft flew a demanding series of pressure distribution experiments at Langley Field. Specially strengthened by Boeing to withstand extreme maneuvers, it became one of the few flying machines to enter the Langley inventory by purchase, rather than by loan agreement.

100. Equipped with computerized (analog) stability augmentation, the X-15 (seen here being checked by the ground crew) both benefited from the system and experienced problems as a result of it.

Over the years, the Langley researchers had evolved a systematic, incremental process of testing vehicles, and through it they became familiar with the effects of pressure distribution in a host of combat maneuvers. No less than Doolittle, the NACA pilots pushed the sturdy PW-9 and the other machines to their limits in order to assemble the broadest possible data.[10] By the early 1930s, a stream of NACA technical reports—written principally under the able hand of a young Langley engineer named Richard Rhode—announced to the world aeronautics community the design principles required to avoid catastrophic failures in the skies. For the first time, aircraft designers could rely on relatively simple formulae to counteract the force of air pressure on various parts of the airframe. For example, the NACA research enabled Rhode to explain to an audience of Air Service engineers in Dayton the theoretical complexities and the practical limitations of tail loading on combat machines. Then, in the first published report on the subject, he admonished military pilots with

regard to the inadequacy of the existing design standards for pursuit airplanes. As these and other reports rolled out of Langley—dissecting safe loading on each of the aircraft's structural members—the fame of Rhode and the NACA spread. Not only builders of military aircraft but also airline manufacturers and governmental entities responsible for air safety peppered Langley with requests for additional data.[11] By the time Rhode and his colleagues completed their last paper on the subject, this important flight research project had transformed the aircraft design process and, in many cases, the designs themselves.

Nearly fifty years after the pressure-distribution work began, a project pursued at NASA Dryden made an impact on aeronautics (and, indirectly, on space flight) at least equal to that of the research begun by Jimmy Doolittle.[12] The concept of uniting computers with aircraft actually has a lineage that stretches back long before the advent of true digital-controlled flight. As early as the late 1940s, the manufacturer of the famed B-49 flying wing equipped it with an analog stability augmentor designed to make it fly as if it possessed a tail surface. During the 1950s, increasing numbers of military aircraft featured "black boxes," that is, analog computers that modified the pilot's stick inputs on planes prone to instability in flight. The X-15 hypersonic aircraft also flew with computer aids. Because of the X-15's immense flight envelope, its designers at North American Aviation subcontracted with the computer firm Honeywell for an analog adaptive control system activated during the re-entry (glide) phase of the flights. On pitch, yaw, and roll maneuvers the Honeywell system shared authority with the pilots, but not without some occasional lapses.[13] Still, hypersonic flight proved that pilots required the assistance of electronic controls (fig. 100).

The leap from atmospheric flight to space flight during the 1960s also occasioned a technological jump forward in the relationship between computing and flight. Astronauts flying space vehicles at enormous rates of speed required piloting assistance available by no other means than computers. Analog computers had already

shown their worth for stability augmentation aboard the lifting bodies; for pilot control over the utterly non-aerodynamic Lunar Landing Research Vehicles (flown by Neil Armstrong and the other astronauts to simulate lunar landings on earth); and on board the Gemini capsules (fig. 101). Meanwhile, several NASA engineers studied the feasibility of positioning a digital computer between the cockpit of an aircraft and its many control surfaces.

This radical concept materialized in a project proposed by Dryden engineers Calvin Jarvis and Melvin Burke. They envisioned a full-scale flight research program that would marry a sophisticated analog computer to a nimble military aircraft. They took this idea to NASA headquarters, where they met with the Deputy Associate Administrator for Aeronautics Neil Armstrong, fresh from his walk on the moon. Armstrong objected. Why an analog system, when a fledgling digital one had just brought him to the lunar surface and back? Burke and Jarvis admitted with some embarrassment that they failed to think of the Apollo computer system, and immediately embraced Armstrong's suggestion (fig. 102).

101. Neil Armstrong, shown here undergoing lunar landing practice, experienced both the X-15 and the Lunar Landing Research Vehicle analog computer systems during his career as a research pilot at NASA Dryden.

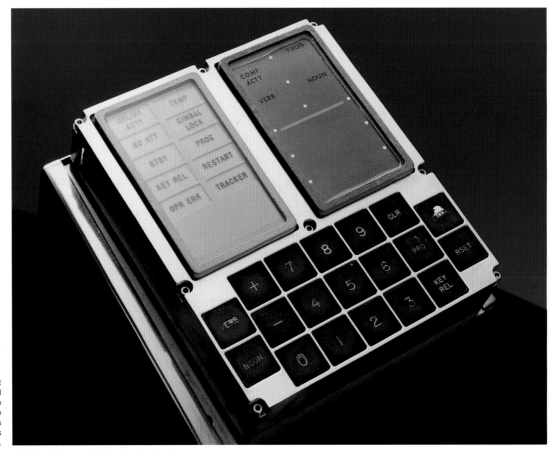

102. The computer used in the first digital fly-by-wire flights. Neil Armstrong recommended it to the NASA Dryden engineers in charge of the digital fly-by-wire project; it is similar to the one he used in his flight to the moon.

103. Digital Fly-By-Wire's pilot and plane: Gary Krier and a Navy F-8C Crusader. Technicians at Dryden Flight Research Center removed all of the mechanical linkages that were common to flight until that time and replaced them with computers and electrical wire.

Led by Jarvis and a young Dryden researcher named Ken Szalai, the Digital Fly-By-Wire (DFBW) group selected an available Navy F-8C Crusader to be the testbed, a supersonic aircraft known for stability throughout its flying range. They then made a more difficult and a more fundamental decision. Jarvis and Szalai decided to fly all of the experiments with no mechanical redundancy. That is, the technicians assigned to the project received instructions to remove all cables, actuators, and push and pull rods running from the cockpit to the flight control surfaces of the aircraft. For the first time in the history of aeronautics, a computer and electrical wire— rather than mechanical linkages—would operate the ailerons, flaps, rudder, and horizontal stabilizer.

Not unexpectedly, the adaptation of the Apollo computer assumed prime importance to this project. The Dryden researchers and the software engineers at the Charles Stark Draper Laboratory in Cambridge, Massachusetts (who had helped program the Apollo system) collaborated on the painstaking, line-by-line process of writing the computer software, with Dryden supplying the control laws and software specifications and Draper the programming. The two parties, at opposite ends of the country, spent whole days on the telephone, attempting to iron out discrepancies. These problems often came to light during the endless hours devoted to "flying" simulators programmed to approximate the handling qualities of the F-8 during actual flight. Perhaps the most persistent and worrisome deficiency manifested itself when the Apollo computer repeatedly shut down as it detected programming errors. The answer lay in redundant computer systems, each machine operating independently of the others, but linked to one another to assume control sequentially if one of the systems shut down.

Finally, NASA research pilot Gary Krier made the world's first all-digital aircraft flight in May 1972 (fig. 103). Although it was an undoubted milestone in the annals of aeronautics, the event did not occur without difficulties. Krier experienced a disquieting one-second pause in which the flight control surfaces failed to respond to his commands. This interruption occurred during the brief interval between primary computer shutdown and assumption of authority by its backup. In addition, because the computer

processed data in digital segments, the act of moving the aircraft control stick resulted in a feeling of repetitive bumps, rather than a fluid motion, as the aircraft responded.[14] Both of these problems diminished to some extent with the introduction of a new generation of computers, designated the AP-101 by their manufacturer, International Business Machines (IBM). The first data processors specially designed for aircraft, the AP-101s failed repeatedly in simulated flights. But when first incorporated into a real flight, on the F-8 in August 1976, the loss of one had absolutely no impact on pilot sensation; the aircraft continued to handle normally as one of the redundant AP-101s took command.

Although the problem of delayed responsiveness to stick inputs never completely disappeared from DFBW flying, the flight research on the F-8 during the 1970s radically affected the design of future aircraft. Taking advantage of the reduction in weight (owing to the absence of mechanical linkages) and the more accurate control of DFBW, such military machines as the Navy F/A-18 Hornet, the USAF F-16, and the Air Force's F-22 Advanced Tactical Fighter all embraced digital flight. The civil sector responded, too, first in the Airbus product line, then during the 1990s in the Boeing 777. But more important, a number of aircraft took to the skies with designs that would have been unthinkable and aerodynamically impossible in the era before DFBW, including the F-117 Stealth Fighter, the B-2 Bomber, and the backwards-winged X-29 Demonstrator. This fluid technology, which began in aeronautics as analog computing, developed in space flight as an on-board digital system, and returned to aircraft as DFBW, also had a pivotal application aboard the Space Shuttle orbiter. Like these other vehicles, the orbiter depended on digital flight controls.[15] In the future, digital computing seemed likely to rewrite the history of aeronautics and spaceflight once again. An on-going flight research project, known as the Intelligent Flight Control System, held the promise of enabling severely damaged aircraft to fly safely, thus transfiguring both military and civil aviation (fig. 104).

EPILOGUE

In the conception of any physical structure—a bridge, an automobile, a highway, or a skyscraper—the intentions of the designer represent perhaps the most elusive part of the

104. (following pages) The Shuttle orbiter during unpowered approach and landing tests at NASA Dryden in 1977. While computer technology was initially transferred from spacecraft to aircraft, it ultimately returned to space as Dryden engineers applied the lessons and software experience of digital fly-by-wire to the orbiter.

process of creation. Oftentimes, the underlying intellectual or artistic influences affecting these individuals may be unclear even to them. But the impact of flight research on the design of aircraft (and spacecraft) has usually been transparent, in part because the method of discovery is so unmistakable. Moreover, the prohibitive cost of fabricating aircraft and the unacceptable consequences of catastrophic failure demanded a full and conscious understanding of the underlying phenomena. Flight research offered the advantage of uncovering these phenomena in a clear and persuasive manner.

The case of the supercritical wing illustrates the point. Conceived by NASA Langley's brilliant aerodynamicist Richard Whitcomb, the supercritical wing essentially turned conventional airfoil design on its head (figs. 105, 106). Interested in finding a shape that would penetrate the transonic range of speed more efficiently than did existing ones, Whitcomb conducted a series of wind tunnel tests. The results suggested that facing the flat side of the wing up and the curved side down—rather than the other way round, which has been customary since the start of aeronautics—resulted in a wing that diminished the transonic shock wave and reduced boundary

105. Richard Whitcomb, the Langley researcher who discovered the supercritical wing, next to a model of his concept. Whitcomb is standing in one of the Dryden hangars during the wing's flight research phase; behind him is the modified F-8C Crusader flown during the flight research program.

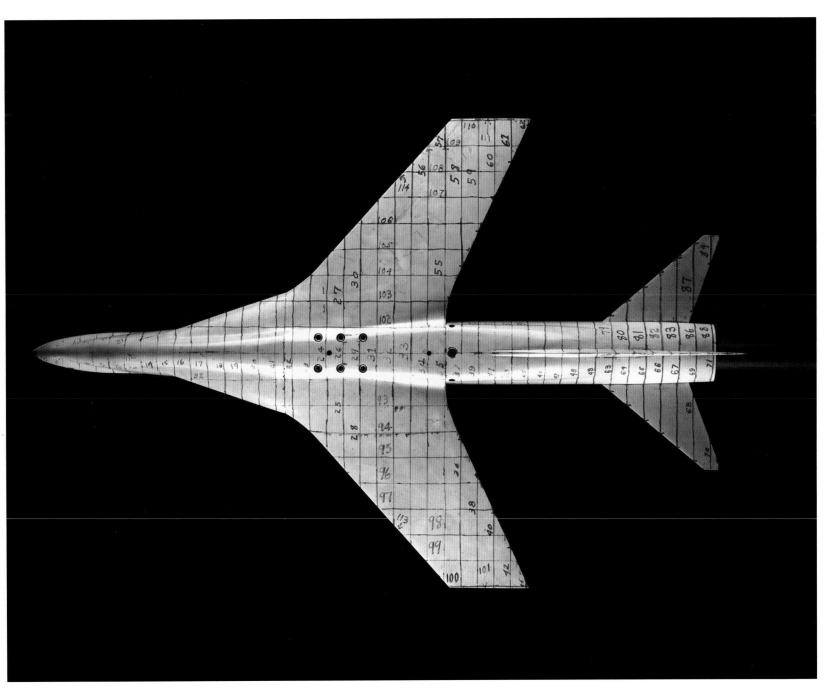

106. F-8 Supercritical wing research airplane model, c. 1968. This model of a modified F-8 was used for supercritical wing research. It displays several of the innovations used to achieve supersonic speeds, including the Area Rule (evident in the pinching of the fuselage) and the supercritical airfoil, both developed at NASA by Richard Whitcomb. This model was specifically used to study the supercritical wing at speeds near Mach 1. The markings on this model represent reference measurements, used to determine the cross-sectional area and model volume. In recognition of his work on the Area Rule Richard Whitcomb was awarded the prestigious Collier Trophy in 1954.

layer separation. Whitcomb predicted a 5% reduction in drag with his design. Yet, even though he had won the prestigious Collier Trophy in 1954 for his milestone discovery of area rule design (supplanting bullet-shaped fuselages with ones narrowed at the wings), his bosses at Langley insisted wind tunnel data alone failed to prove the case for the supercritical wing. Indeed, as Langley's Director of Aeronautics Laurence Loftin remarked pointedly, "This thing is so different from anything we've done before that nobody's going to touch it with a ten-foot pole without somebody going out and flying it."[16] Thus, the supercritical wing, like countless other

innovations, required tangible proof that it would perform as predicted before industry would consider its adoption.

Only flight research could fill this essential role. Moreover, in carrying out their research practitioners often discovered—as they had in the M2-F1's first flight—deficiencies or surprising phenomena not suspected until the moment an aircraft went aloft. Since the first generation of modern flight began at Kitty Hawk, flight research has given designers the confidence to pursue new concepts that have ultimately transfigured not only civil and military aviation but space flight as well.

107. A waverider model,
c. 1980–85. This model was
created to determine the pressure
distributions on the windward and
leeward sides of the vehicle—note
the pressure taps on the left-hand
side of the model (the little holes
at the end of the dark lines). The
waverider concept was based on
the theory that a vehicle could ride
its own shockwave, or bow-wave,
at hypersonic speeds if the shape
of the vehicle were designed to
match the shockwave angle. The
vehicle would have been able to
ride on its own bow-wave, much
like a boat on water.

Notes

1. The NACA's, and later NASA's, research facility on Edwards Air Force
Base has been known by a number of names: the Muroc Flight Test Unit,
the High-Speed Flight Research Station, the High-Speed Flight Station, the
Flight Research Center, the Hugh L. Dryden Flight Research Facility, and
the Hugh L. Dryden Flight Research Center. For reasons of simplicity it is
referred to as NASA Dryden or as Dryden in this essay.

2. R. Dale Reed with Darlene Lister, *Wingless Flight: The Lifting Body Story*,
Washington, D.C. (NASA SP-4220) 1997, p. 36.

3. For more on the lifting body see *ibid.*, pp. 23–36; see also Milton O.
Thompson and Curtis Peebles, *Flying without Wings: NASA Lifting Bodies
and the Birth of the Space Shuttle*, Washington and London (Smithsonian
Institution Press) 1999, especially pp. 69–70.

4. For technical and historical definitions of flight research see, respectively,
Kenneth J. Szalai, *Role of Research Aircraft in Technology Development*,
Edwards, Calif. (NASA Technical Memorandum 85913) 1984; and Michael
H. Gorn, "Introduction," in *Expanding the Envelope: Flight Research at
NACA and NASA*, Lexington, Ky. (The University Press of Kentucky) 2001.

5. The history of flying qualities research in the United States is covered in
Walter G. Vincenti, *What Engineers Know and How They Know It: Analytical
Studies from Aeronautical History*, Baltimore (The Johns Hopkins University
Press) 1990, chapter 3; and also in Gorn, *op. cit.* (n. 4), chapter 3.

6. "Report for Research Aircraft Projects Panel of Research Activities of
NACA High-Speed Flight Research Station for the Year 1951," pp. 1 and
37; "Meeting Minutes of Interlaboratory Research Airplane Projects Panel,"
February 4–5, 1952, pp. 4–5; "Report for Research Airplane Projects Panel
of Research Activities of NACA High-Speed Flight Research Station for the
Year 1952," pp. 2–3 and 51–61; Hartley Soule to NACA Headquarters,
January 5, 1953, "Agenda Items for Research Airplane Projects Panel
Meeting on January 14 and 15, 1953," p. 5; "Addendum, Minutes of
Meeting of Interlaboratory Research Airplane Projects Panel," 4–5 February
1954, p. 3; "Report for Research Airplane Projects Panel of Research
Activities of NACA High-Speed Flight Research Station for the Year 1954,"
pp. 2 and 57. All of the above manuscripts are filed in the Hallion Papers,
NASA Dryden Flight Research Center Historical Reference Collection.

7. C.A. Syvertson, "Aircraft Without Wings," *Science Journal*, December
1968, pp. 46–48; Reed with Lister, *op. cit.* (n. 2), pp. 9–15; Gorn, *op. cit.*
(n. 4), pp. 258–74 and 283–88. See also Thompson and Peebles, *op. cit.*
(n. 3), chapters 4–8.

8. *Against the Wind: Ninety Years of Flight Test in the Miami Valley*, Dayton,
Oh. (Aeronautical Systems Center) 1994, pp. 10 and 11.

9. James H. Doolittle, *Accelerations in Flight*, NACA Technical Report 203,
Washington, D.C. (National Advisory Committee for Aeronautics) 1925,
pp. 373–88.

10. Gorn, *op. cit.* (n. 4), pp. 57–75.

11. *Ibid.*, pp. 75–96.

12. See James E. Tomayko, *Computers Take Flight: A History of NASA's
Pioneering Digital Fly-By-Wire Project*, Washington, D.C. (NASA SP-4224)
2000.

13. J.P. Sutherland, "Fly-By-Wire Flight Control Systems," Society of
Automotive Engineers (SAE)-18, Aerospace Vehicle Flight Control
Committee, Boston, Mass., August 10, 1987, pp. 1–5, Hallion Papers,
DFRC Historical Reference Collection; Duane McRuer, interview with Lane
Wallace, Hawthorne, Calif., August 31, 1995, p. 2, DFRC Historical
Reference Collection; William Elliott, *The Development of Fly-By-Wire Flight
Control*, Dayton, Oh. (Air Force Materiel Command History Office) 1996,
p. 13; Robert A. Tremant, *Operational Experiences and Characteristics of
the X-15 Flight Control System*, Washington, D.C. (NASA Technical Note
D-1402) 1962, pp. 1–2 and 12; Milton O. Thompson and James R. Welsh,
"Flight Test Experience with Adaptive Control Systems," Summary of
Papers Presented at AGARD Guidance and Control and Flight Mechanics
Panels, Oslo, Norway, September 3–5, 1968, p. 5, DFRC Historical
Reference Collection.

14. Gary Krier, telephone interview with Michael Gorn, March 11, 1999,
DFRC Historical Reference Collection; Kenneth J. Szalai and Calvin R.
Jarvis, interview with Lane Wallace, August 30, 1995, pp. 2–7, 13, 19, and
22, DFRC Historical Reference Collection; Tomayko, *op. cit.* (n. 12), p. 31;
Richard P. Hallion, *On the Frontier: Flight Research at Dryden, 1946–1981*,
Washington, D.C. (NASA SP-4303) 1984, pp. 217–18; Elliott, *op. cit.* (n. 13),
p. 13; Ronald "Joe" Wilson, interview with Michael Gorn, April 11, 1997,
pp. 31–32, DFRC Historical Reference Collection.

15. Gorn, *op. cit.* (n. 4), pp. 325–28.

16. The flight research program at Dryden proved Richard Whitcomb's
calculations to be essentially correct, and the supercritical wing has
become a standard feature on airliners the world over. Richard T.
Whitcomb, "Research on Methods for Reducing the Aerodynamic Drag at
Transonic Speeds," The Inaugural Eastman Jacobs Lecture, NASA Langley
Research Center, Hampton, Va., November 14, 1994, pp. 5–6, DFRC
Historical Reference Collection; Richard T. Whitcomb, "The State of
Technology Before the F-8 Supercritical Wing," in *Proceedings of the F-8
Digital Fly-By-Wire and Supercritical Wing First Flight's Twentieth
Anniversary Celebration at NASA Dryden Flight Research Center, Edwards,
California, 27 May 1992*, Washington, D.C. (NASA Conference Publication
3256) 1996, vol. 1, p. 85 (quoted passage); meeting notes, Laurence Loftin,
"Discussions of Supercritical Wing Research Airplane," March 21, 1967,
Hallion Papers, DFRC Historical Reference Collection; Warren C. Wetmore,
"New Design for Transonic Wing to be Tested on Modified F-8," *Aviation
Week and Space Technology*, February 17, 1969, p. 23; Tom Kelly,
interview with Richard Hallion (handwritten notes), April 24, 1978, Hallion
Papers, DFRC Historical Reference Collection.

CHAPTER 06

Dennis R. Jenkins

DESIGNING FOR THE EDGE OF SPACE AND BEYOND

109. Every aspect of a vehicle is tested before it is used in flight. Here a pair of Reaction Motors XLR11 rocket engines in an X-15 are test-fired at Edwards Air Force Base. The X-15 had been designed to use a Reaction Motors XLR99 rocket engine that produced 57,000 pounds (12,815 N) of thrust, but that engine was not available when the flight program began. In its place, the XLR11s produced only 16,000 lbf (3600 N), but could power the airplane over Mach 3 and 100,000 feet (30,500 meters). Eventually, 30 flights would use the XLR11s; 169 would use the XLR99.

108. The X-15 vehicle, considered to be one of the most successful flight research programs, is seen during one of its last flights with a dummy ramjet engine experiment mounted under its aft fuselage. A joint NASA, US Air Force, and US Navy program, the X-15 completed 199 flights between 1959 and 1968. It advanced numerous technologies in high-speed flight and in crew systems.

110. Vehicle design often progresses during the life of a program. Here are three 1/15-scale wind tunnel models of the North American X-15 research airplane. The model at the left shows the basic configuration of the airplane as it was delivered. The middle model shows the "advanced" X-15A-2 configuration that was applied to the second airplane after it was damaged in an emergency landing. This configuration went on to set an absolute speed record for winged aircraft of Mach 6.70 (4520 miles/7274 km per hour) with Major William J. "Pete" Knight at the controls in 1967. This record would stand until the return of the Space Shuttle Columbia from its first orbital mission in 1981. The model at the right was a proposed delta-wing modification that was never flown. These models are currently displayed at the AFFTC Museum at Edwards Air Force Base.

Despite the fanciful predictions of Eugen Sänger, Wernher von Braun, and a wealth of science fiction novelists, it was not until the mid-1950s that the first piloted spacecraft design was undertaken in earnest.[1] It was the height of the Cold War, and the paranoia that swept the country and the military had resulted in the largest arms race the world had ever seen. In aviation the desire was to go higher, faster, and farther than ever before. In response to a need for basic research into the ever-increasing speeds and altitudes, the National Advisory Committee on Aeronautics (NACA) began preliminary research into a piloted vehicle that could exceed five times the speed of sound. The research was felt necessary to support both unmanned missile programs and the eventual development of hypersonic combat aircraft. Interestingly, the group of researchers that took the lead in developing the concept (led by John V. Becker) at the NACA's Langley Laboratory added a new wrinkle—they wanted to be able to leave the sensible atmosphere for a few minutes in order to gain a preliminary understanding of space flight.[2] At the time it was generally felt that piloted space flight would not take place until the turn of the century, although contemporary science fiction—a genre that enjoyed a resurgence of popularity in the mid-1950s—usually showed it coming much earlier. In fact, many serious researchers believed that the group at Langley should remove the "space leap" from their concept for a hypersonic research airplane.[3] However, the basic designs for a very high speed airplane and for one capable of short excursions outside the atmosphere were not radically different, so the capability remained.

THE RACE FOR SPACE
The conceptual design proposed by Langley was eventually to become the North American X-15 research airplane (fig. 108). Formal development was begun in 1955 under the leadership of Harrison Storms and Charles Fletz, with a first flight expected in early 1959. The airplane was the first to broach many of the technologies needed of a spacecraft. The science of aerothermodynamics played a major role, and the entire structure was designed around the problem of surviving the tremendous heat generated in high-speed flight, or during re-entry from outside the atmosphere. The designers worried about how to control the vehicle after the air became

too thin for the normal aerodynamic controls; the now-familiar reaction control system was the result. The pilot would of course need protection from low pressure and radiation, and the David Clark Company, with significant imput from test pilot Scott Crossfield, was responsible for developing the first workable full-pressure suit.

Since air data would no longer be available to determine the position of the airplane, a crude "stable platform" was developed that was the precursor to modern inertial systems. The vehicle would require extraordinary power, and Reaction Motors Incorporated responded with the first large throttleable and restartable rocket engine, the XLR99 (fig. 109).[4] The effects of high acceleration on human beings needed to be better understood, so the X-15 pilots became guinea pigs in a fascinating research program that took place in a human centrifuge at the Naval Air Development Center, Johnsville, Pennsylvania.[5] Everything from the hydraulic oil to the instrumentation in the cockpit would need to be specially developed for this new flight regime (fig. 110).

It must be remembered that the X-15 was being designed before any man-made object had been sent into space, and the engineers are preparing for a step-by-step assault on the unknown. Each piece of the puzzle would be analysed before the pilots made the next step. It was a logical, if time-consuming, way to "fly" into space.[6]

Everything changed on October 4, 1957, when the Soviet Union launched Sputnik (later called Sputnik 1), the first artificial earth-orbiting satellite. The effect on the United States was profound. The Soviet achievement embarrassed American scientific and technological prestige, and the satellite was widely regarded as a threat to national security. Robert Gilruth, who would go on to head Project Mercury, wrote, "I can recall watching the sunlight reflecting off the Sputnik 1 carrier rocket as it passed over my home on the Chesapeake Bay, Virginia. It put a new sense of value and urgency on the things we had been doing."[7]

The US Army and Navy renewed their efforts to launch an unmanned satellite, while Congress began proceedings that eventually led to the creation of the National Aeronautics and Space Administration (NASA). The Air Force began serious consideration of a minimal program to launch an American into space, an effort that was eventually transferred to NASA and became Project Mercury.[8] Although North American, alongside several other, companies proposed vehicles that would fly into space, none of the rocket boosters then in existence could launch such a payload.[9] It was decided instead to use a simple ballistic capsule design, and the human occupant became more of a passenger than a pilot (fig. 112).

Meanwhile, in California, work continued on the X-15 with a renewed emphasis. It should be noted that wind tunnel testing was—as it still is—an inexact science. For example, in order to observe shockwave patterns across the X-15's shape, small (3–4 inch / 7.5–10 cm) models were "flown" in the hypervelocity free-flight facility at the NASA Ames Research Center; the goal was to shoot the model out of a gun at tremendous speeds.[10] The models were made out of cast aluminum, cast bronze, or various plastics, and were actually fairly fragile. As often as not, what researchers saw were pieces of X-15 models flying down the range sideways. Fortunately, sufficient numbers of the models remained intact to acquire meaningful data (fig. 111).

One of the major problems was how to build an airplane capable of withstanding equilibrium temperatures of 2000° F (1093° C). At the time there were no known insulators that could protect the airframe, so researchers at the NACA developed a "hot structure" that could absorb the heat. A special stainless-steel alloy called Inconel X was the primary structural material, but it was only capable of absorbing about 1200° F (649° C) when used in a traditional structure.[11] To overcome this, the X-15 structure was exceedingly heavy, with very thick skins that allowed it to absorb more heat without melting. Unfortunately, as it grew hotter it wanted to buckle and warp. So the engineers devised ways that allowed each piece of the structure to expand in at least one dimension without resistance, eliminating most of the distortion. It was an imaginative solution to the problem, but one that required exhaustive attention to detail.[12]

Data from the X-15 centrifuge program would later be used as a basis for much of the physiological conditioning required of the Mercury astronauts. The efforts of the International Latex Corporation and the David Clark Company led to the development of the Mercury and Gemini space suits, respectively. The Gemini suit was heavily based on the suit that had been developed for the X-15. The common perception of a space

111. Austrian physicist Ernst Mach took the first photographs of supersonic shockwaves using a technique called a shadowgraph. In 1877 Mach presented a paper to the Academy of Sciences in Vienna where he showed a shadowgraph of a bullet moving at supersonic speeds; the bow and trailing edge shockwaves were clearly visible. Mach was also the first to assign a numerical value to the ratio between the speed of a solid object passing through a gas and the speed of sound through the same gas. In his honor, supersonic velocities are expressed as "Mach numbers." Conceptually, this Schlieren photo of an X-15 is not much different than the shadowgraphs taken by Ernst Mach almost one hundred years earlier. Some of the shockwaves seen here are being generated by protrusions on the other side of the airplane.

112. The launch of Sputnik by the Soviet Union in 1957 had a profound effect on the US. Among other things, it greatly accelerated the low-level effort that had been started to place a human into orbit. In late 1957 and 1958 the X-15 was briefly considered as a possible vehicle. Here, Wernher von Braun is seen discussing a North American proposal to use the X-15 in space. Left to right: (aid to von Braun), von Braun, Fred Payne, Alvin S. White, Howard Evans, Raymond Rice, and Harrison Storms. Although the concept of flying into space was very appealing to most engineers, none of the available boosters had the capacity to place a winged object into orbit. The ballistic capsules were significantly lighter and were (barely) within the capability of the boosters, so the ballistic capsule concept was ultimately chosen for Project Mercury.

113. Models are frequently tested in many different wind tunnels around the country. Each tunnel may have specialized capabilities that are of use to the program, or it simply may be a way to meet schedules. Here is an X-15 model in a hypersonic tunnel at the US Air Force Arnold Engineering Development Center (AEDC) near Tullahoma, Tennessee. The X-15 was also extensively tested in NASA wind tunnels at Ames, Langley, and the Jet Propulsion Laboratory, as well as at MIT, the University of Washington, and the Southern California Cooperative Tunnel (a tunnel owned and operated by the various aerospace contractors located around Los Angeles).

suit has it covered in silver lamé. Interestingly, the original David Clark full-pressure suits were a dirty khaki color because that was the natural color of the fabric used to construct them. When a photographer first arrived at the Navy laboratory that had been funding the development of early full-pressure suits, he was decidedly unimpressed with the khaki-colored cloth garment. To give him something a bit flashier to photograph an old hard-shell ILC suit was rolled out—and it was this that later appeared on the cover of a national magazine, despite the fact that it had already been deemed unworkable and obsolete. When the first of the X-15 suits was being manufactured, Scott Crossfield suggested to David Clark that perhaps it should be covered with a silver lamé that Crossfield had seen lying on a table in the factory. There was no technical reason, but both Crossfield and Clark thought it would look better and would be more photogenic. Clark decided that using black boots and gloves would provide a good contrast, although the black gloves would

later prove to be too hot. Thus, mainly to appeal to the camera's eye, the modern perception of a space suit was born.[13]

TESTING THE X-15

To a large degree, the art and science of designing the X-15 was much the same as that employed for the airplanes that had come before it; wind tunnel models made of steel were used for the initial tests (fig. 113). But the X-15 marked the beginning of a more systematic approach, as was necessary in order to prepare for flight into higher temperatures and higher dynamic pressures than those encountered before. Much of the science gathered by the X-15 during its flight program would be used to verify the accuracy of various analytical theories and predictive modeling techniques used during its development. The methods used to gather the data were varied. Pressure and temperature sensors built into the structure of the airplane took precise readings at hundreds of points. A special thermally sensitive paint was frequently

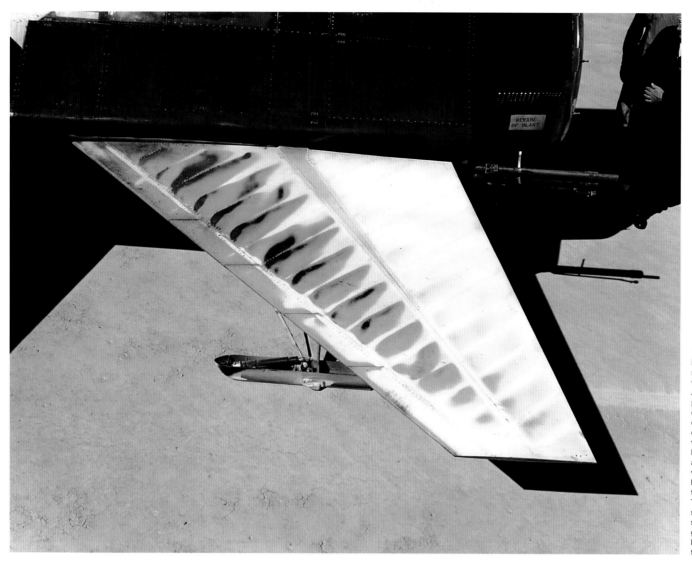

114. Data is often gathered in interesting ways. This horizontal stabilizer on an X-15 was painted with a thermally sensitive paint. As the airplane got hotter, the painted surfaces turned different colors, and then maintained those colors once the airplane cooled down. This allowed researchers visually to see the heat gained by various surfaces. The white areas here were relatively cool, while the brownish areas were hot. The internal structure under the skin can clearly be seen. Thermocouplers (temperature sensors) were also installed in the skin and underlying structure, giving more precise measurements, but the visual effect is sometimes the easiest to understand.

used to coat various portions of the airplane, turning nearly psychedelic colors as it went through various heat ranges.

When the flight program began, most of the data reduction was done by hand; as the program ended ten years later, digital computers were in routine use to accomplish the same tasks (fig. 114).[14]

Perhaps even more important than the hardware, the operational aspects of the X-15 would shape nearly every future flight research program to some degree (fig. 115). The X-15 was the first airplane that was extensively simulated, both for pilot training and as a research tool. The simulator used a complete set of actual flight hardware electronics and hydraulics, coupled to an analog computer (fig. 116). The most complex parts of each mission could be simulated in real-time, with the pilot flying from a detailed fixed-

base cockpit. It was not uncommon for twenty hours of simulator time to go into each eight- to ten-minute flight. The simulator could also predict heating values for locations on the airframe in real-time based on how the pilot was flying the mission. The X-15 simulator that was the most advanced at performing these temperature analyses would later be used for the same purpose on Space Shuttle.[15]

Late in the flight program one of the X-15s was rebuilt to enable it to fly even faster—Mach 8 was the goal (figs. 117, 118). It was obvious that the innovative hot structure could not absorb the heat load at this velocity, so a special ablative coating was developed that would protect the airframe by sacrificing (charring) itself. This material was largely sprayed over the airframe like a thick paint, and as it cured it turned a rather intriguing shade of pink. Fortunately, it needed to be weatherproofed with a white overcoat, saving

115. Flight planning used to be much simpler. Here Scott Crossfield shows the flight plan for an early X-15 flight. The names of the chase pilots are at the left top corner of the chalkboard, while the flight path and all critical events are shown near Scott's hand. Although the tools may seem simplistic, the X-15 program successfully planned and executed a breathtaking flight expansion program that took piloted aircraft from just over Mach 3 to almost Mach 7 in just seven years.

116. Contrary to popular belief, most aircraft simulators are not particularly glamorous to look at. This is the X-15 simulator at the Flight Research Center in 1961. The X-15 program was one of the first to make extensive use of simulation, not only for pilot training but also as an engineering tool. This simulator was a "fixed-base" simulator, meaning it did not move. However, it was equipped with a fully functional cockpit, and a complete set of electronics and hydraulic systems that were controlled by an analog computer. The computer was "mechanized" (programmed) to behave exactly like an X-15; the model was changed continually, based on the results of wind tunnel and flight tests.

117. After a hard landing, which actually broke the vehicle in two, the second X-15 vehicle was modified for even higher speeds and altitudes. Propellant drop tanks and an elongated fuselage were added to increase on-board fuel storage, and new ablative insulation was applied (not shown) to counteract the high temperatures of flight. Pete Knight would take this vehicle to a speed of Mach 6.7. This photo was taken at the airplane's rollout ceremony in Los Angeles.

118. The relative size of the two external propellant tanks on the modified X-15A-2 is readily apparent in this unusual view. The propellants in these tanks essentially doubled the amount of time the XLR99 rocket engine could burn.

the test pilots the embarrassment of flying a pink airplane. At the time it was expected that some sort of ablative coating would be used on Space Shuttle, so there was a great deal of interest from many designers. The ablative proved to be time-consuming to apply and refurbish, although not really any worse than initially predicted. It has been widely reported that the ablator failed on its two full-up test flights, but in reality it performed mostly as expected; the publicized failures were very localized and were largely due to problems unrelated to the ablator itself. Nevertheless, given the time required to apply it to the X-15, it was hard to imagine how it would work on a vehicle many times as large (fig. 119).[16]

The designers of the X-15 knew that the actual shape and construction of the airplane would likely never be used for another vehicle, but the lessons learned and data gathered were believed to be applicable to any winged spacecraft (fig. 120). Certainly the data proved useful to the stillborn X-20 Dyna-Soar program that was the heir apparent. But the space program got sidetracked by the move toward developing ballistic capsules, and it would be almost two decades before another winged re-entry vehicle was developed. While some of the data from X-15 would be used by the engineers on Gemini and Apollo, it was Space Shuttle that would be the ultimate beneficiary.

FROM BLACKBIRD TO SPACE SHUTTLE

Just a few years after the development of the X-15 began, a talented group led by the legendary Kelly Johnson at the Lockheed Skunk Works initiated the classified CIA program that led to the SR-71 Blackbird reconnaissance aircraft. Unlike the rocket-powered X-15, the Blackbirds would use turbojet engines; for thirty years the Mach 3+ Blackbirds would be unchallenged as the fastest air-breathing aircraft in the world. Much like the X-15, the group at Skunk Works would need to develop almost every aspect of the airplane from scratch. Although less than half as fast as the X-15, the Blackbird presented a different problem. An X-15 flight lasted eight to ten minutes; the Blackbird cruised at Mach 3+ for an hour at a time. Sustained heating—instead of peak heating—was the concern. Johnson's design proved up to the task, and added to the legend of the Skunk Works development group. Interestingly, the X-15

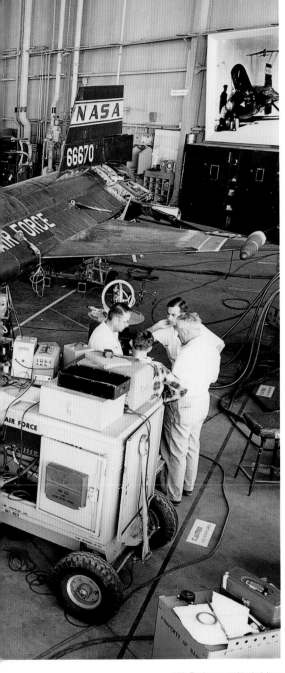

119. Engineers and technicians check out the X-15-1 prior to a flight. The open compartment behind the cockpit is where the research instrumentation was located. By this point, the airplane had been equipped with small wing-tip pods to carry additional experiments outside the flow disturbance created by the airplane. The vehicle is pictured in the main hangar at NASA Dryden.

120. The main hangar at NASA Dryden, Hangar 4802, pictured in 1966. At that time the main flight research programs were the lifting bodies and the X-15. Visible in the photograph are three of the lifting bodies, the three X-15s, NASA's C-47, and a few other research aircraft.

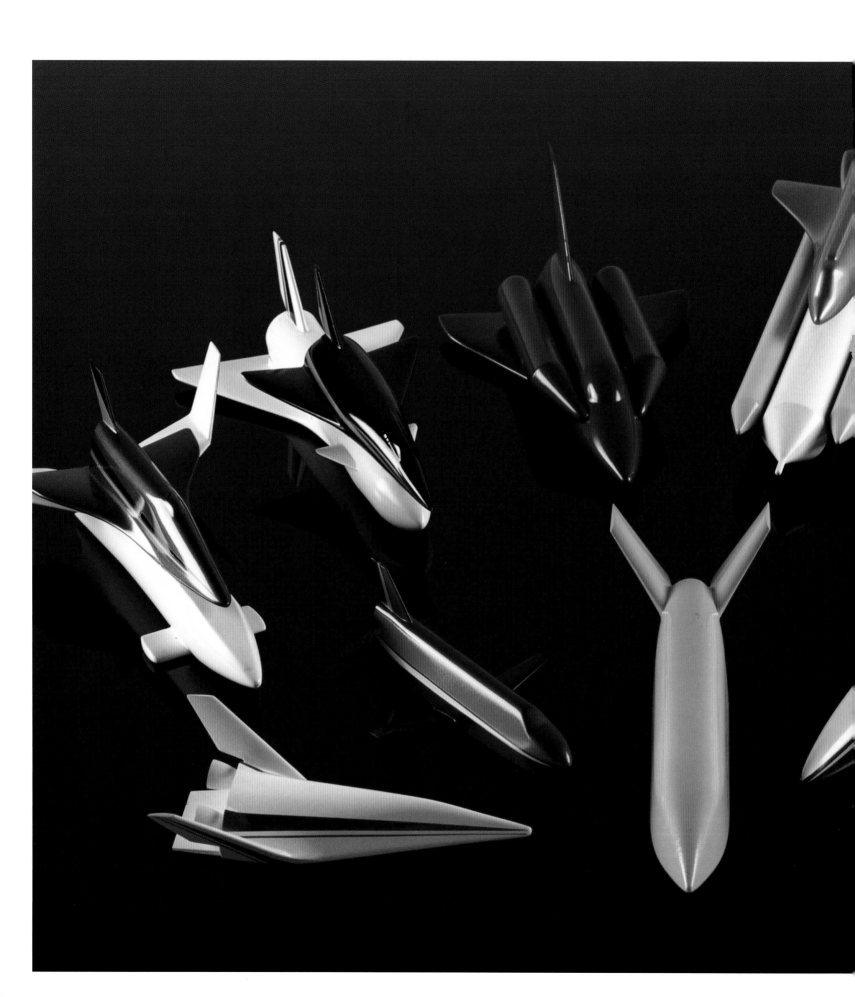

122. A wind tunnel model of the Space Shuttle orbiter, c. 1970–78. The era of Space Shuttle began on April 12, 1981, when the Space Shuttle Columbia lifted off from the pad at the Kennedy Space Center on its first orbital flight. Since that time, the fleet of four Space Shuttles has made well over one hundred orbital flights. Space Shuttle has been called the most complex system ever flown, with over a million parts. The program has been a success despite the explosion of the Challenger on January 28, 1986, and the loss of the Columbia on February 1, 2003 during re-entry. The Space Shuttle orbiter is similar in size to a Boeing 737 or DC-9. NASA has consistently upgraded the shuttles to fly well into the first decades of this century; meanwhile engineers are striving to develop a replacement.

121. A collection of wind tunnel models showing the evolution of the Space Shuttle vehicle during development, 1969–72 (the models are not to scale). The idea of an integrated launch and re-entry vehicle was initially proposed while the Saturn program was flying men to the moon in 1969. The initial concept for this program was for a fully reusable multistage vehicle, but due mainly to reasons of cost a partially reusable vehicle was developed instead. This is what we know today as the Space Transportation System, or Space Shuttle, with its reusable orbiter, refurbishable solid rocket boosters, and expendable external tank. An early model of the final configuration is shown at top center, while some of the very early fully reusable two-stage vehicles are at the left.

123. The maiden flight of the Space Shuttle Columbia, April 12, 1981. The Space Transportation System (STS) was the culmination of over a decade and a half of design and development. Space Shuttle is composed of three major elements. The solid rocket boosters (SRBs) provide the majority of the thrust necessary to fly into space. The external tank (ET) contains the propellants used by Space Shuttle's main engines (SSME) on the back of the orbiter, which provide the remaining thrust. The airplan-like orbiter contains the crew and payload. The vehicle launched vertically, as rockets had done in the past, while the orbiter would re-enter the atmosphere and fly back as a glider.

actually contributed to the spying capability of the Blackbird; it had tested special high-temperature camera windows at very high speeds.[17]

When the development of Space Shuttle began in the late 1960s, the X-15 was still flying. By the time Columbia made its first flight in 1981, the X-15s had been in museums for thirteen years (fig. 123). Space Shuttle faced many of the problems encountered by the X-15, but multiplied by a thousand percent. This did not seem seriously to discourage the designers, however (figs. 121, 122), and some of the early concepts were truly fanciful. Huge two-stage, fully reusable systems employed a large hypersonic carrier aircraft to launch a slightly smaller (but still large) orbiter. The fact that the only hypersonic aircraft ever to fly was the 30,000 pound (13,600 kg) X-15 did not seem to worry those proposing slightly faster carriers that weighed 3,000,000 pounds.[18] The correlation between the predictive theories used to build the X-15 and actual flight results, plus the new computer-based simulations developed out of the X-15 experience, gave designers more confidence than they probably deserved (fig. 124).

124. (below) New designs are tested in a variety of ways. Here a ¹⁄₁₀-scale model of an early Space Shuttle concept is being drop tested from an Army CH-54 Skycrane helicopter. The model was 13 feet (4 meters) long and weighed 600 pounds (272 kg). After the model was released from the helicopter, engineers could assess the low-speed handling characteristics of the design. A parachute would lower it the final few hundred feet to the ground. The same design was also tested in a variety of higher-speed wind tunnels in order to provide data over its entire operating range.

The difficulties encountered with the ablative coating on the X-15 underscored the need to find another way to protect Space Shuttle from the heating environment encountered both in high-speed flight during the boost phase, and from re-entry temperatures coming back from orbit (figs. 125–27). Many ideas were proposed, including the use of thin shields of exotic alloys held a few inches away from the main airframe, and advanced versions of the ablator used on the X-15. An unlikely candidate was a special tile made from—of all things—sand and air. In the end the tile won the day, and given its absolute simplicity, it is truly a marvel. The center of a tile that measures a few inches across can be glowing red-hot, yet the edges can comfortably be held by an unprotected human hand. It turned out that the tiles had a new set of problems, including the question of how to attach them to the vehicle, but they nevertheless represent one of the more remarkable material developments that contributed to the progress of the Space Shuttle project.[19]

The X-15 had proven that it was possible for a human successfully to pilot a spacecraft during the boost and re-entry phases. It was realized that this was probably not the ultimate answer to the question of manned space flight, however, and even late in the X-15 flight program special systems were devised that made the piloting tasks easier—adaptive control systems, predictive energy management displays, and the like. With the tremendous increase in computer power available to the designers of Space Shuttle, much of this process was automated, although many of the techniques were simple extensions of what had been learned earlier. In fact, the only spacecraft to be manually flown back from orbit was the Space Shuttle Columbia (on November 14, 1981), by Joe Engle—one of the twelve men who had flown the X-15. Space Shuttle became the first vehicle that was software-centric, although by today's standards it has remarkably few lines of code. Still, Space Shuttle features four computers that run identical software and vote on every critical decision to make sure that they do not make a mistake. A fifth computer runs in the background with an entirely different, albeit very limited, set of software in case the programmers made a catastrophic mistake in the primary set.

125. The flow field around a model of the Space Shuttle orbiter is visible at Mach 20 and 35 degrees angle of attack. An electron beam illuminates the flow around the orbiter model in the 22 inch (56 cm) leg of the NASA Langley Research Center Hypersonic Helium Facility. The variations in color and brightness graphically depict the flow around the vehicle.

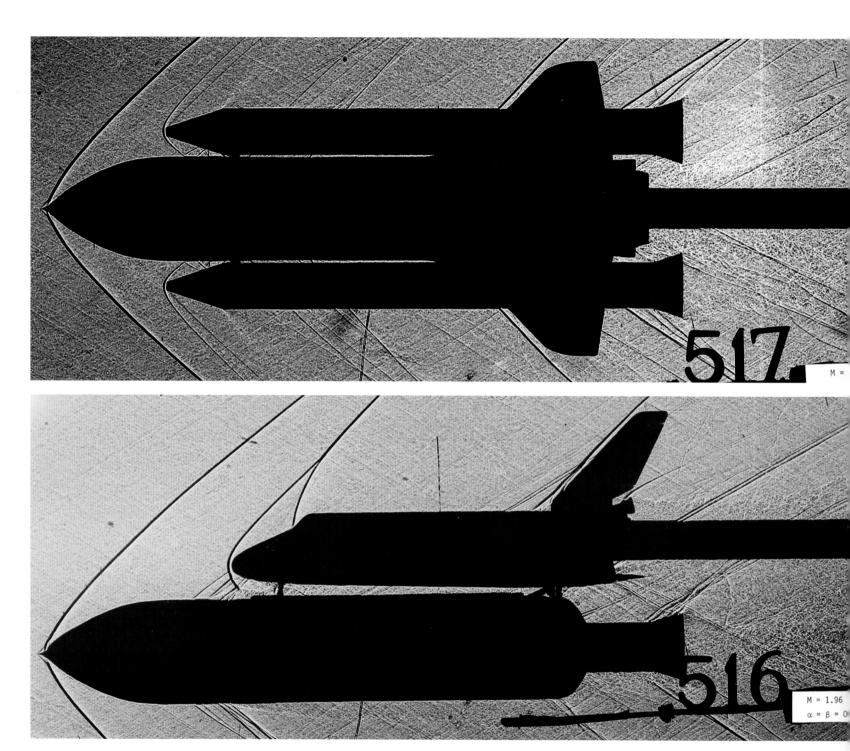

517.

M =

516

M = 1.96
α = β = 0

126. Two Spark shadowgraphs,
taken at Mach 1.96, highlight the
shockwaves forming over a model
of the Space Shuttle launch vehicle
during wind tunnel testing at NASA
Marshall Space Flight Center.

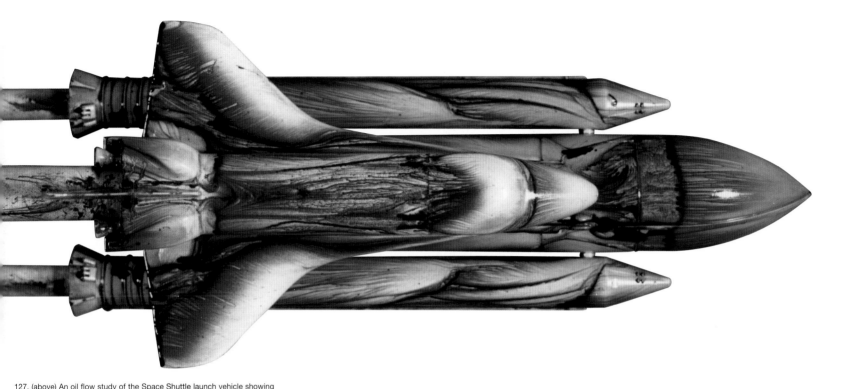

127. (above) An oil flow study of the Space Shuttle launch vehicle showing the localized flow patterns over the vehicle when the vehicle is traveling at Mach 1.96. A mixture of oil and paint is applied to the model before testing in the wind tunnel. During testing, the airflow "blows" the mixture over the model. When the wind tunnel test is completed, light colors or lack of paint denote high pressure regions, while dark blue colors or heavy concentrations of paint show low pressure or stagnation regions.

128. (below) A wind tunnel model of the Space Shuttle launch vehicle, 1970s. During launch and ascent Space Shuttle consists of the orbiter (the plane that will glide back to earth), the external tank that holds the liquid oxygen and hydrogen that will fuel the main engines on the orbiter, and the two solid rocket boosters that look like candles. The three main engines on the orbiter and the two solid rocket boosters power the orbiter into orbit, along with its crew and payload. The solid rocket boosters burn for about two minutes then separate from the external tank, to be recovered for reuse after their landing on water. Once the main engines are shut down, the external tank is jettisoned and burns up upon re-entry into the Earth's atmosphere over the Indian Ocean. Over a million hours of wind tunnel testing were competed on Space Shuttle before its first flight.

The computers run everything. In the aircraft that came before Space Shuttle, the pilot's control stick was connected to a series of cables or push-rods that activated the controls directly, or commanded a set of hydraulic actuators directly. In Space Shuttle the control stick is connected to the computer. One of the X-15s had already tried a similar technique, called the MH-96 adaptive flight control system. But in the X-15 the mechanical linkages still existed, and the pilot could turn off the adaptive system and fly home the old-fashioned way if necessary. This option is not available to the pilots on shuttle. It was a new game, and NASA was extremely conservative. The software onboard the orbiters is the most extensively tested software ever produced. In a time when mission-critical software could be written, documented, and tested for $50 per line, NASA paid over $1000 per line. It was not wasteful spending; a later analysis showed that the software was essentially "error free"—good news to the seven astronauts that depended upon it for their lives.[20]

Space Shuttle marked another radical departure for the aerospace industry. During the seventy years prior to its development, engineers had generally attempted to make their airplanes as stable as possible. Various stability augmentation systems had been developed to take care of slight instabilities in certain flight regimes, but

in general the aircraft could be flown with those systems inoperable. The X-15 had demonstrated that it was not really possible to build a vehicle that would be truly stable over the wide range of speeds and altitudes necessary to fly into space and return. If the adaptive control system on the X-15 failed during re-entry from extremely high altitudes, it was unlikely the pilot could control the airplane. Space Shuttle faced a similar problem, only to an even greater extent. The orbiter ended up being unstable throughout much of its flight regime—it is up to the computer to manipulate the control systems in such a manner that the vehicle is flyable (figs. 129, 130). If the computers

fail, the orbiter stops flying. Of course, if the computers fail, the pilot has no direct control linkage anyway, so it really does not matter. Similar fly-by-wire systems are now commonplace on most military aircraft, many commercial transports, and some high-end automobiles.[21]

FUTURE TECHNOLOGIES

Space Shuttle proved it was possible to build a large, reusable re-entry vehicle. Given the technology available to its designers it is a remarkable feat, and one that has worked extremely well for more than twenty years

129. A ⅓-scale model of the final Space Shuttle orbiter configuration was tested in the 40 × 80 Foot (12 × 24 meter) Wind Tunnel at the Ames Research Center. The model was 43.9 feet (13.38 meters) long, and was used for low-speed data in support of the approach and landing tests that would be conducted using the Space Shuttle Enterprise at Edwards Air Force Base, California.

130. Spacecraft meets aircraft nose to nose: the Space Shuttle Enterprise riding atop its carrier aircraft, a NASA wide-body 747. The Space Shuttle orbiter Enterprise was launched five times during 1977 from a 747 carrier aircraft in unpowered glide flights. These flights verified the gliding and landing characteristics of the new orbiter.

131. A NASA Marshall Space Flight Center test engineer examines a proposed modification to Space Shuttle, a liquid boost module. Thousands of hours of wind tunnel tests were completed throughout the United States on the shuttle and its components. Any proposed modifications would be wind-tunnel tested before being implemented. Space Shuttle is not alone in requiring this degree of testing— any new aircraft design undergoes extensive testing using various models at numerous wind tunnels before the actual vehicle is built.

(fig. 131). But much like the X-15 before it, it is unlikely that any future vehicle will make direct use of Space Shuttle hardware or technology.[22] What will be used are much the same lessons that were learned from the X-15. Simulation, analysis, predictive theory, training, procedures, and techniques have all been advanced considerably with the lessons learned from Space Shuttle. But an entire new generation of composite and alloy materials, and incredible leaps in computer power, will let the designers of the next vehicle do things that were not even dreamed of during the development of Space Shuttle.

This is not to say it will be easy, or even easier. Already a variety of "replacement" vehicles—real or imagined—have fallen by the wayside. The X-33 and X-34 technology demonstrators pushed "state-of-the-art" design a little too far in some areas, and were discarded before their first flights.[23] A new generation of research vehicles is emerging, however, continuing in the tradition of the X-15 and Space Shuttle. The X-43 hypersonic air-breathing test vehicle picks up on a technology that was being explored just as the X-15 flight program ended. Being able to use relatively uncomplicated air-breathing engines instead of immensely complex and costly rocket engines will provide new options for future designers of spacecraft and fast transport aircraft. Computers may have replaced the research pilots, but the basic philosophy of exploring new frontiers of flight is the same as it was fifty years ago when the X-15 was first conceived. The future beckons.

Notes

1. The German scientist Eugen Sänger proposed a suborbital vehicle as early as 1928, and during World War II some conceptual development of the Silverbird "Amerika Bomber" actually took place. This vehicle would be almost 100 feet (30.5 meters) long, weigh 200,000 pounds (90,700 kg), and be launched by a rocket sled. It would fly at 13,500 miles (21,725 km) per hour at the edge of space, eventually covering 14,500 miles (23,335 km) before landing. See, for example, Eugen Sänger and Irene Bredt, "The Silver Bird Story: A Memoir," in R. Cargill Hall (ed.), *Essays of the History of Rocketry and Astronautics: Proceedings of the Third Through Sixth History Symposia of the International Academy of Astronautics*, Washington, D.C. (NASA) 1977; and Eugen Sänger, *Rocket Flight Engineering*, from the German *Raketenflugtechnik*, translated by NASA as TT F-223, Washington, D.C. (NASA) 1965.

2. For an overview of early winged re-entry vehicles, see Dennis R. Jenkins, *Space Shuttle: The History of the National Space Transportation System–The First 100 Missions*, North Branch, Minn. (Specialty Press) 2001, pp. 1–47.

3. For an overview of the hypersonic program (which led to the development of the X-15 research airplane), see Dennis R. Jenkins and Tony Landis, *Hypersonic: The Story of the North American X-15*, North Branch, Minn. (Specialty Press) 2003. The most frequently quoted X-15 history was originally written in 1959 by Robert S. Houston, a historian at the Air Force Wright Air Development Center. This narrative, unsurprisingly, centered on the early Air Force involvement in the program, and concentrated—as is normal for most Air Force histories—mostly on the program management aspects rather than the technology. Dr. Richard P. Hallion, now the Chief Historian for the US Air Force, updated Houston's history in 1987 as part of volume II of *The Hypersonic Revolution*, a collection of papers published by the Aeronautical Systems Office at Wright-Patterson Air Force Base. Hallion added coverage of the last nine years of the program, drawing mainly from his own *On the Frontier: Flight Research at Dryden, 1946–1981*, Washington, D.C. (NASA) 1984, and a paper written in 1977 by the Air Force Academy's Ronald G. Boston, titled "Outline of the X-15's Contributions to Aerospace Technology."

4. The Reaction Motors XLR99 proved to be very difficult to develop, and lagged seriously behind its original schedule. It also accurately predicted the cost overruns that would be associated with the development of most large rocket engines. The original engine estimate was about $6,000,000. The letter contract was for $9,961,000; by the time the final engine contract had been signed, the estimate had already risen to $10,160,030, plus an additional $614,000 in fees. At the end of the financial year 1958 engine costs had risen to over $38,000,000, and expenditures in the financial year 1959 brought the cost to $59,323,000. Estimated engine costs for 1960 were $9,050,000—almost as much as the total X-15 program estimate of 1955. As of June 1959, the engine costs were $68,373,000—more than five times the original estimate for the entire X-15 program and almost a seven-fold increase over the costs contemplated when the engine contract was signed. Each of the ten "production" engines cost just over $1 million.

5. The centrifuge at the Naval Air Development Center at Johnsville, Pennsylvania, was an extraordinary device. A 4000 horsepower (2980 kW) electric motor rotated a 50 foot (15 meter) arm with a gondola on its end and could produce accelerations of up to forty times the force of gravity. The X-15 pilots were the first group of aviators to undergo training in the device, although the Mercury astronauts that followed them received a great deal more publicity for essentially the same training.

6. The amount of uncertainty that surrounded some aspects of space flight is quite astonishing. The reaction control system that was proposed to control the altitude of the vehicle outside the atmosphere was an interesting case. The system operated by converting hydrogen peroxide to superheated steam by decomposing it over a silver catalyst. This steam was then exhausted out of thrusters in the nose and wingtips, essentially becoming small rocket engines. The uncertainty came when it was time to determine how much hydrogen peroxide to carry; some engineers thought that 50 or 60 gallons (190 to 228 liters) would be required; others thought as little as 3 gallons (11.4 liters) was sufficient. In the end, it was discovered that about 2 gallons (7.6 liters) were normally used on a high-altitude flight.

7. Robert R. Gilruth, "From Wallops Island to Project Mercury, 1945–1958: A Memoir," in R. Cargill Hall (ed.), *History of Rocketry and Astronautics*, American Astronautical Society History Series, vol. 7, part 2, San Diego (American Astronautical Society) 1986, p. 462.

8. The Air Force effort was called Project 7969, soon to evolve into Project MISS—Man-in-Space-Soonest. By the end of 1957 the Air Force had received at least ten serious proposals, which were evaluated by a joint Air Force–NACA team during a conference held at Wright Field on January 29–31, 1958.

9. Lockheed and Martin each proposed a blunt re-entry vehicle like that used later for the Discoverer/KH-4/Corona film recovery capsules. McDonnell proposed a shape similar to the later Soyuz. Avco, Goodyear, and Convair proposed spheres, like that used for Vostok. Bell, North American, Republic, and Northrop all proposed winged vehicles. See Loyd S. Swenson Jr., James M. Grimwood, and Charles C. Alexander, *This New Ocean: A History of Project Mercury*, Washington, D.C. (NASA SP-4201) 1966, pp. 75–132.

10. Dale L. Compton, "Welcome," in *Proceedings of the X-15 30th Anniversary Celebration*, Dryden Flight Research Facility, Edwards, California, June 8, 1989 (NASA CP-3105), p. 3. The free-flight tunnel at Ames had been conceived by H. Julian Allen and opened in 1949 at a cost of about $20,000. It had a test section 18 feet (5.6 meters) long, 1 foot (30 cm) wide, and 2 feet (60 cm) high. By forcing a draft through the tunnel at a speed of about Mach 3 and firing a model projectile upstream, velocities of up to Mach 18 could be simulated. Schlieren cameras were set up at seven locations along the test section, three on the side and four on the top, to make shadowgraphs that showed the airflow over the models. The facility proved to be an important tool not only for the X-15 but also for Project Mercury.

11. Inconel X ® is a temperature-resistant alloy; its name is a registered trademark of Huntington Alloy Products Division, International Nickel Company, Huntington, West Virginia. Inconel X is 72.5% nickel, 15% chromium, and 1% columbium, with iron making up most of the balance.

12. Unlike many airplanes where the structure is made as solid as possible, much of the X-15 structure was loosely connected. Detailed thermal analyses revealed that large temperature differences would develop between the upper and lower wing skin during the pull-up portions of certain trajectories. This unequal heating would result in intolerable thermal stresses in a conventional structural design. To solve this problem, wing shear members were devised that did not offer any resistance to unequal expansion of the wing skins. The wing thus was essentially free to deform both spanwise and chordwise with asymmetrical heating. It was also discovered that differential heating of the wing's leading edge produced changes in the natural torsional frequency of the wing unless some sort of flexible expansion joint was incorporated in its design. The hot leading edge expanded faster than the remaining structure, introducing a compression that destabilized the section as a whole and reduced its torsional stiffness. To negate these phenomena, the leading edge was segmented and flexibly mounted in an attempt to reduce thermally induced buckling and bending.

13. Story in A. Scott Crossfield, *Always Another Dawn: The Story of a Rocket Test Pilot*, North Stratford, NH (Ayer Company Publishers) 1960, pp. 241–42 and 254–55. The development of pressure suits is a fascinating story and centers largely around three individuals: Scott Crossfield, David Clark at the David Clark Company, and Russell Colley at B.F. Goodrich. Crossfield's autobiography contains some interesting insights into the process. A general overview is presented in Jenkins and Landis 2003.

14. The X-15 was the first airplane to be extensively instrumented from the beginning (several other X-planes had instrumentation added during their flight programs). It was also the first airplane to make extensive use of telemetry, allowing the engineers on the ground to see the instrumentation in real-time, and the X-15 team pioneered many of the concepts required to use the data. This technique became routine during the manned space program and directly contributed to the success of Mercury, Gemini, and Apollo.

15. It should be noted that these simulations were run on analog computers, not digital ones. Although digital computers existed at the time, they were generally very slow and could not produce the simulations in real-time. During the course of the X-15 program, the simulator was upgraded to a hybrid configuration where most of the real-time simulation was still accomplished on the analog computers, but a digital unit was added to perform the more complex heating studies. The simulator that supported Space Shuttle was located at the Air Force Flight Test Center at Edwards Air Force Base and had originally been designed to support the Boeing X-20 Dyna-Soar program. After that program was cancelled it was reprogrammed to support the advanced X-15A-2. During this reprogramming the engineers added a "generic" capability to simulate almost any winged aircraft, hence its adaptability to support Shuttle. Unfortunately, many of the advances in the early simulators have never been written about. Although it concentrates only on the NASA experience at the Flight Research Center, Gene L. Waltman, *Black Magic and Gremlins: Analog Flight Simulations at NASA's Flight Research Center*, Washington, D.C. (NASA SP-2000-4520) tells a good story about the technology and techniques available at the time. For a great overview of the flight program told by a pilot who was deeply involved in it, see Milton O. Thompson, *At The Edge of Space: The X-15 Flight Program*, Washington, D.C. (Smithsonian Institution Press) 1992.

16. Ablators were used to protect all of the manned ballistic capsules and most missile warheads and other re-entry vehicles. They work very well in these one-time applications. The primary difficulty lies in trying to reuse the ablator. By its very nature, an ablator "chars" and sacrifices itself in order to insulate the structure it is protecting. It therefore needs to be refurbished prior to another use. Ablators tend to be very time-consuming to apply, and make it very difficult to work on the vehicle after they have been applied (it covers access doors, etc.). Nevertheless, during the early conceptual design process for Space Shuttle, ablators were usually thought to be the best answer. See Jenkins, *op. cit.* (n. 2), various pages but especially pp. 156–60.

17. The Blackbird development effort was one of the most successful "black" (*i.e.*, secret) programs ever undertaken, and resulted in a truly magnificent machine. See Dennis R. Jenkins, *Lockheed SR-71/YF-12 Blackbirds*, vol. 10 in the WarbirdTech series, North Branch, Minn. (Specialty Press) 1997, for a look specifically at the Blackbirds. For more on the Skunk Works—arguably one of the most fascinating design organizations in the history of aviation— see Jay Miller, *Lockheed Skunk Works: The First 50 Years*, Hinckley, UK (Midland Publishing) 1995.

18. The sheer size of some of the early fully reusable two-stage concepts is mind boggling. For instance, in 1970 North American proposed a booster that was 269 feet (82 meters) long with a wingspan of 143 feet (43.6 meters), and which used twelve 550,000 lbf (123,640 N) rocket engines to lift an orbiter that was 192 feet (59 meters) long and spanned 126 feet (39 meters). The booster released the orbiter at 7000 miles (11,300 km) per hour—almost twice as fast as the X-15 ever flew. For comparison, the Boeing 747-200 passenger jet was 230 feet (70 meters) long, spanned 195 feet (59 meters), and had a maximum gross weight of 800,000 pounds (362,880 kg). It was powered by four 55,000 lbf (12,364 N) turbofan engines.

19. The tiles are made of low-density, high-purity, 99.8% amorphous silica fibers derived from common sand, ⅛ inch (1 to 2 mm) thick, which are made rigid by ceramic bonding. Because 90% of the tile is void and the remaining 10% is material, the tiles weigh approximately 9 pounds per cubic foot (9 kg per cubic meter). A slurry containing fibers mixed with water is frame-cast to form soft, porous blocks to which a colloidal silica binder solution is added. When it is sintered, a rigid block is produced that is cut into quarters and then machined to the precise dimensions required for individual tiles. HRSI (High-Temperature Reusable Surface Insulation) tiles can vary in thickness from 1 to 5 inches (25 to 127 mm), with the thickness determined by the heat load encountered during re-entry. Generally, the HRSI tiles are thicker at the forward areas of the orbiter and thinner toward the aft end. The HRSI tiles are nominally 6 by 6 inch (152 by 152 mm) squares, but vary in size and shape in the closeout areas. The black tiles are coated on the top and sides with a mixture of powdered tetrasilicide and borosilicate glass with a liquid carrier. This material is sprayed on the tile to coating thicknesses of 16 to 18 mm. The coated tiles then are placed in an oven and heated to a temperature of 2300° F (1260° C). This results in a black, waterproof, glossy coating that has a surface emittance of 0.85 and a solar absorptance of about 0.85. After the ceramic coating heating process, the remaining silica fibers are treated with a silicon resin to provide bulk waterproofing. As an aside, an HRSI tile taken from a 2300° F oven can be immersed in cold water without damage. Also, surface heat dissipates so quickly that an uncoated tile can be held by its edges with an ungloved hand seconds after removal from the oven while its interior still glows red. For more information, see Jenkins, *op. cit.* (n. 2), pp. 395–402.

20. *Ibid.*, pp. 234–35; Edward J. Joyce, "Is Error-Free Software Achievable?" *Datamation*, February 15, 1989, pp. 53–56; B.G. Kolkhorst and A.J. Macina, "Developing Error-Free Software," *IEEE AES Magazine*, November 1988, pp. 25–31.

21. For the early history of fly-by-wire development, see James E. Tomayko, *Computers Take Flight: A History of NASA's Pioneering Digital Fly-by-Wire Project*, Washington, D.C. (NASA SP-2000-4224) 2000. The Airbus series of commercial transports is almost completely fly-by-wire. On automobiles, the trend began on some Mercedes models and Corvettes when the throttle became a transducer instead of being directly hooked to the engine. Some Mercedes and BMWs now offer brakes that are operated electronically instead of via the normal hydraulics.

22. Unlike most aircraft—indeed, most other machines—spacecraft seldom seem to take the direct hardware lessons of their predecessors. This is most likely because we simply do not develop new spacecraft often enough, and technology changes very drastically in between development efforts. But many of the other lessons are applicable, almost regardless of what technology is actually used to build the vehicle.

23. The X-33 was a subscale prototype of the proposed VentureStar™ reusable launch vehicle. The program had a host of problems, largely political and managerial, but its major technological flaw was the desire to use composite tanks to contain its cryogenic propellants (liquid oxygen and liquid hydrogen). The hydrogen tanks failed on several occasions during proof testing, leading to the program becoming significantly over budget and behind schedule. For an overview of the X-33 program, see Dennis R. Jenkins, *Lockheed Secret Projects: Inside the Skunk Works*, St. Paul, Minn. (MBI Publishing Company) 2001, pp. 95–114.

CHAPTER 07

Anthony M. Springer

THE FUTURE OF FLIGHT

We entered the nineteenth century at 6 mph, the speed
of a horse-drawn cart. We entered the twentieth at 60, the
speed of a steam locomotive. We entered the twenty-first
at 600, the speed of an intercontinental jet airliner. Might
we not enter the twenty-second at 6000, the speed of a
hypersonic transport?

Richard P. Hallion, Chief Historian, US Air Force[1]

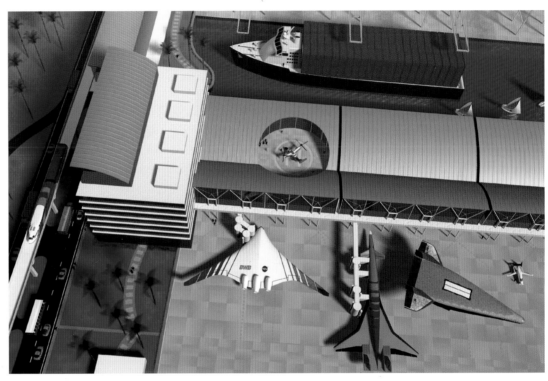

133. (right) Supersonic transports,
hypersonic planes, runway-
independent aircraft, and small
personal aircraft are among the
possibilities for the aircraft of
the future.

132. Possible aircraft of the future
range from the Canard Rotor and
small personal jet aircraft to
supersonic business jets. The
Canard Rotor wing combines the
possibility of vertical takeoff and
landing with very high cruise
speeds.

What does the future hold for aviation and space
flight (fig. 133)? Many people ask why time and
money is spent on such things. A common
thought is: I can fly anywhere I want in an
airplane; rockets launch payloads into orbit every
week; what is there still to be discovered? At the
turn of the last century, skeptics thought there
was nothing new to discover in physics—yet
since then the theory of relativity, atomic energy,
and the digital age have changed our lives. We
are now at a crossroads in the history of flight.
The basic theories of flight and rocketry were
discovered during the last century, but great
leaps are yet to come. The worlds of air and
space are converging. In the future there may be
no distinction between aircraft and spacecraft;
rather, we may have vehicles that can travel in
both worlds. What are the possibilities in travel
for the years to come? A world of point-to-point
travel via "air cars," as seen in cartoons such
as *The Jetsons* (fig. 132)? Large aircraft carrying
a thousand passengers over ever-greater
distances? High-speed transcontinental aircraft

traveling at supersonic or hypersonic speeds,
capable of journeying from Los Angeles to Tokyo
in a few short hours? Or, a reversion to a more
primitive time, that of endless lines, delayed
flights, and a captive public with few options?
Regardless of speed or capacity, the vehicles
of the future are likely to be safer, more
environmentally friendly, and more economical
to operate than current models, as a result of
advances in technology.

The past may hold the key to the future of flight.
What we have come to know as flight has been
around a very short time. The flight of humans in
any form began a mere 220 years ago, with the
first balloon launch in 1783. Controlled powered
flight (1903) has existed for a century; liquid
fueled rocketry (1926) for a little over three-
quarters of a century; jet aircraft (1939) and
supersonic flight (1947) for a little over half a
century; rockets into space (1958) for less than
fifty years; and human visits to the moon (1969)
for only thirty-five years. From the early fragile

biplanes made of wood and cloth, lacking the power and structure to carry out their mission, planes have evolved into the sleek metallic and composite aircraft of today.

Over the course of its short history the aerospace industry has witnessed some dramatic changes. Its evolution has been punctuated by revolutionary ideas that set a new course; these ideas then evolved until the next revolutionary idea or technology was introduced. One might picture the industry as a series of sloping steps. Each vertical rise or step is a revolutionary change; each sloping plateau is the evolutionary period of the industry. The first revolutionary change would be powered flight, followed by metal construction, the jet engine, and high speed flight. What will be the next revolutionary change? In terms of commercial aircraft we have been on our current "sloping plateau" since the 1950s, with the advent of commercial jet transports. The designs have evolved, but no revolutionary change has taken place. (This is not to say that the commercial airliners have not evolved over the last fifty years.) On the military side we have had significant advances in performance and capabilities, specifically in fighter aircraft and uncrewed aerial vehicles (UAVs). UAVs give us a glimpse of what the future may hold. The next step may not be just in the vehicles but in the air transportation system itself.

As we look at the future of aviation and space flight we must consider not only the vehicles but also the whole system. The vehicles we see and fly in are but a small part of a much larger network. An understanding of the factors that influence this network and its growth will be crucial to the creation of less congested airports. The savings in construction and operation costs brought about by advances in technology may bring a myriad of additional benefits, from increased safety to lower fuel consumption and cheaper air fares.

WHERE ARE WE TODAY?

Over the last thirty years, advances in aircraft technology have reduced aircraft noise and accident rates by a factor of ten, despite a three-fold increase in flight operations. Today, air travel is one of the safest forms of transportation: the likely loss of life in a commercial aircraft is about one in a million.[2] Fuel consumption and cost have been reduced by 50%. Since 1978 revenue passenger miles (RPM) have increased by 190%, while ton-miles have increased by 289% and are increasing at a rate of 10% annually.

In the United States there are about 5400 airports. Of these, most passengers pass though only 1%, and only 10% are used to any degree (i.e. commercially). The air traffic system in the United States handles 63 million aircraft operations—take-offs and landings—every year, which entails 544 million passengers traveling over 537 billion passenger miles. This equates to the distance from the earth to beyond the edge of our solar system. It is estimated that flight delays, both on the ground and in the air, cost passengers and the industry $4.5 billion annually, the equivalent of a 7% tax on each ticket. With no change to the current system, where would this lead us? It is projected that by 2007 delays will cost industry and passengers $13.8 billion every year, rising to $47.9 billion by 2017. Although airplanes are the safest form of transportation today, if current accident rates are combined with the expected increase in the number of flights, an accident a week can be expected within the next two decades.[3]

Examine the average airplane flight from Washington, D.C. to Chicago, Illinois. For this pair of cities the traveler has a choice of three major airports from which to leave Washington, and in Chicago there are two options. Regardless of the pair chosen, the journey time will be about the same. The goal is to go from downtown Washington to downtown Chicago. You leave work and walk to the subway station (fifteen minutes), taking the subway to the airport (twenty minutes). At the airport you proceed though check-in and security, and walk to your gate at the correct terminal (thirty minutes; this neglects the current one hour or more additional security requirement). After boarding (twenty minutes) the aircraft taxies to the runway, takes off, and flies to Chicago. Assuming there are no weather problems and there is no congestion in the air routes, the flight will take an hour and twenty minutes. Upon landing, the aircraft taxies to the assigned gate and the passengers disembark (twenty minutes). Once at the terminal you walk from your gate to baggage claim, collect your

134. An advanced fighter concept featuring the latest technologies.

135. A single-seat F-16XL (NASA 849) joins up with an SR-71A (NASA 844) in preparation for a sonic boom study, one of several undertaken by NASA at the Dryden Flight Research Facility during the 1990s. The F-16XL was used to probe the shockwaves generated by the SR-71 at altitude, while other aircraft and ground sensors were used to record the data at other altitudes and on the ground.

bags, and walk to the Chicago subway (twenty minutes) for a short ride downtown (twenty minutes). A short walk would then finally bring you to your destination (fifteen minutes). The elapsed time of this theoretical trip is about four hours, of which one hour and twenty minutes is spent in the air, an hour and a half at the airport, and one hour going to and from the airport.

What does this sample trip tell us? Using the current system, the passenger spends as much time at the airport as in the air, and almost as much traveling to and from the airport. How might this scenario be improved? The answer lies in

making savings to both the flight time and the ground time.

FLIGHT TIME

Without drastic changes to the aircraft itself, the time spent in flight can only be slightly improved. Current commercial aircraft cruise at about 600 miles (966 km) per hour, or Mach 0.80 to 0.85. With advanced vehicles such as Boeing's proposed Sonic Cruiser, this could be increased to about Mach 0.95. Yet flying at supersonic speeds is not without its problems (fig. 135);[4] any flight at or over the speed of sound results in the generation of shockwaves from the aircraft.

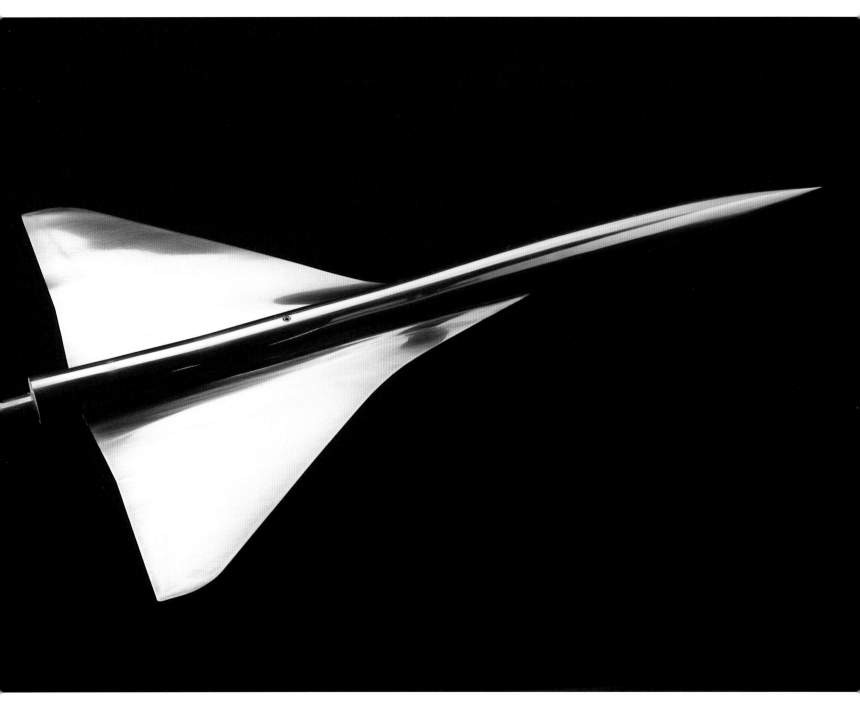

136. This wind tunnel model is one of many designs studied during the High-Speed Civil Transport (HSCT) program. During the late 1980s and early 1990s numerous configurations were studied for a supersonic commercial passenger aircraft. A supersonic aircraft would allow a dramatic decrease in flight time but encountered may technological hurdles, not the least of which was the sonic boom caused by aircraft flying at supersonic speeds.

137. A series of delta-wing wind tunnel models, 1980s. Alexander Lippisch
(1894–1976) was one of the great pioneers of delta-wing aircraft,
publishing the results of his life's work in *The Delta Wing*. He is best
remembered as the creator of record-breaking swept-back-wing Storch
gliders during the 1920s, and as the designer of such famous planes as
the Me-163 Komet (1940-45; the world's first successful production rocket
fighter), and the Convair F-102 and F-106 supersonic delta-wing fighters.
The delta wing is extensively used today for high-speed flight. Both Space
Shuttle and the Anglo-French Concorde, along with many proposed high-
speed or supersonic aircraft, use or propose to use a delta wing.

138. A proposed supersonic airliner capable of sustained Mach 2.4 flight. A supersonic airliner would halve flight times compared with current aircraft. During the 1990s NASA and its industry partners studied technologies that would make aircraft such as these economically feasible and less environmentally damaging.

139. The first supersonic commercial airliner was the Soviet TU-144, which entered service in 1975, beating the British–French Concorde to the market by a year. It soon disappeared from the scene, however, due to a series of crashes and a lack of cost-effectiveness. During the mid-1990s a joint NASA and industry team used the TU-144 to study new supersonic flight technologies.

Filghts In and Out of Dallas - Fort Worth
Color Coded by Altitude
Low Altitude - High Altitude

140. A computer visualization of aircraft flights into Dallas/Fort Worth International Airport during a typical day.

This effect, similar to that of a bullet coming from a gun, leads to "sonic booms"—noises so loud that they will quite literally rattle windows in buildings on the ground. It is for this reason that commercial supersonic flight is not allowed over the continental United States. In any case, at these distances the difference in flight time is negligible when other journey elements are considered, saving only about thirty minutes on a normal five-hour transit time for a cross-country flight. To alleviate the problem of sonic boom, research is currently being conducted into new configurations that could reduce the noise heard on the ground to acceptable levels (figs. 138, 139).

One well-known example of commercial supersonic flight is Concorde's journey from Europe to the United States. Flying at over

Mach 2, Concorde makes the trip in three hours and thirty minutes. By comparison a Boeing 747-400, flying at Mach 0.85, requires about seven hours for the same crossing.[5] The flight from Los Angeles to Tokyo—a distance of 4750 miles (7644 km)—currently takes eleven hours on a standard jet airplane (the return flight is ten hours due to prevailing winds); current proposals for supersonic transport flying at Mach 2.2 would cut the trip to four hours each way. For an advanced hypersonic aircraft traveling at Mach 6 these flight times would be reduced from hours to minutes. The technology is currently under development to make this kind of sustained hypersonic flight feasible. At the present time hypersonic flight is neither commercially viable nor technically possible, but if the technologies currently under development and mentioned in

this chapter are employed, hypersonic flight could occur in our lifetimes.

GROUND TIME

To decrease the total transit time the ground segment time must be examined. It was once thought that the solution to airport congestion was more pavement, but new runways will not solve all the problems. Only a limited number of aircraft can land at a given airport in a given amount of time (fig. 140). In the current system, taxi time and loading/offloading are major factors, along with passenger time to the gate. Changes can be made to deal with these problem areas—such as multiple gates or loading through multiple entrances/exits on the aircraft—but ultimately the

only real way to alleviate them is to redesign the airport. Most airports were conceived dozens of years ago and have expanded many times since. Several are encountering passenger numbers much higher than originally envisioned, and are approaching overload. The expansions that are undertaken are often not optimal, due to economic factors. An airport cannot be closed for the years it would take to replace it with a new, more efficient version designed to meet modern-day traffic loads in terms of both people and aircraft. The solution is usually merely to build additional terminals.

New computer-based tools are under development to expand the capabilities of the

142. One way to avoid the long transit times of the "hub and spoke" system would be a "rent-a-jet"-type arrangement. High-tech jet aircraft could be hired at a local airport, and then flown to the local airport at the desired destination. For obvious reasons such aircraft would need advanced control systems that would allow for simpler operation.

141. The Eclipse 500, one of the new breed of personal jet aircraft that could make possible "point-to-point" air travel. Powered by smaller, more efficient jet engines, and featuring cutting-edge avionics, communication and control systems, these vehicles should pave the way for safer and more efficient flight.

144. A modern-day biplane, the box-wing concept holds the promise of providing greater aerodynamic efficiency in a smaller size aircraft. This proposed military version could serve as a future airborne tanker capable of servicing multiple aircraft simultaneously.

143. The blended-wing body (BWB) concept does away with the traditional tubular body: as the name suggests, the body is "blended" with the wings. The result is improved aircraft efficiency.

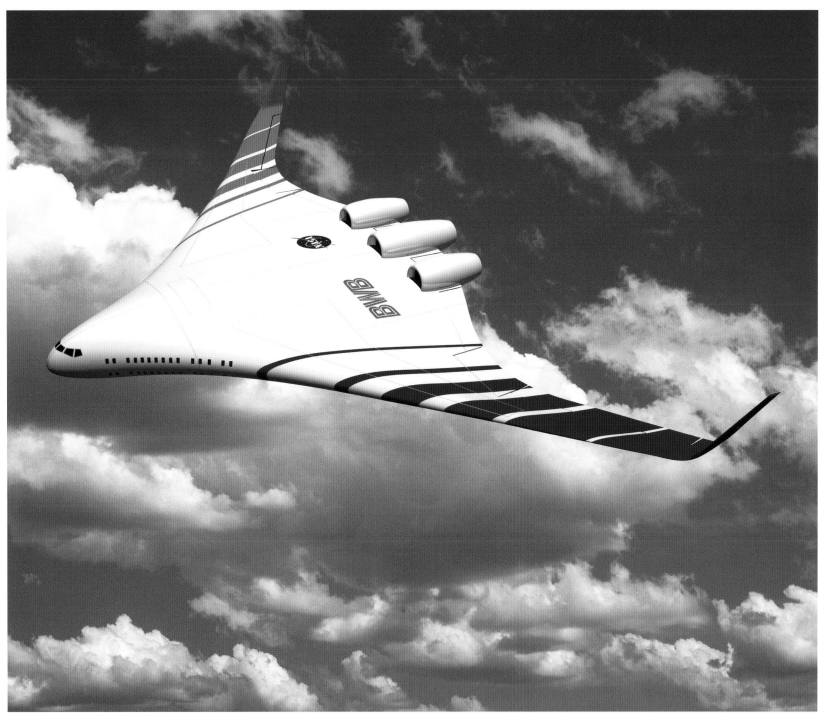

current system. Through better modeling, new processes are being developed to increase the flow of aircraft into and out of airports and to streamline their movements on the ground. These developing technologies may enable more aircraft to take off and land with greater frequency at the same or better safety levels than currently. Other technologies could enable all-weather or increased weather capabilities for the operation of airports. New weather-prediction technologies, along with advanced landing aids, will allow airports to continue to operate fully; at present they have to reduce operations in the case of adverse weather conditions, and the closing of a runway causes delays at not just one airport, but a possible ripple effect through the whole system.

VEHICLES OF THE FUTURE

In years to come we are likely to encounter a range of vehicles specially designed for specific missions. This would be made possible through rapid, low-cost modifications and reliable design. One possible future technology is the "personal" aircraft or air taxi. This vehicle could either be a modified version of the aircraft of today, with improved instrumentation, or a futuristic automated small jet with performance and safety that outstrips that of current vehicles (fig. 141). These new jets could be the rental cars of the future: just pick up a plane at the local airport, follow the simple instructions—keep the dot in the box—and the aircraft almost flies itself (fig. 142). Through "smart systems" on the aircraft, piloting could become as simple as driving a car. The realization of this concept may be in the distant future, but the necessary technologies are being developed today. Technologies are being studied that would enable, among other things, flight in all weather, airborne internet, precise location positioning and advanced terrain avoidance. If these developing technologies could be perfected and brought together, these "smart" aircraft could, in theory, fly themselves.

Other possible future configurations include the blended-wing body, or BWB, and the Stager box wing. Current commercial aircraft all follow the pattern of a tube (the passenger cabin) with appendages (the wings) sticking out of it. In the BWB aircraft the body and the wings are blended together—the concept is not unlike the flying wing of the current B-2 bomber. The BWB would be more efficient than current aircraft—using less

fuel, weighing less, and resulting in an approximate noise reduction of 20 dB—and would be about the same size as the modern-day 747. The Stager box wing is a modern version of the biplane. The use of double "staggered" wings reduces the wingspan, and thus the overall size of the vehicle (figs. 143, 144).

In the mid-1990s the concept of runway-independent aircraft began to be studied for the short-haul civilian market. Hitherto the main focus had been on military applications, the result being the V-22 Osprey (1989). It was back in the early 1960s that NASA, the US Air Force, and the aerospace industry began research into tilt-wing and rotor aircraft as a means to integrate the best points of both the airplane and the helicopter. The helicopter can land just about anywhere and can hover over a single spot, but has poor performance and slow speeds over long distances; whereas the airplane is more fuel efficient and capable of greater speed over long distances, but requires a runway to land and cannot hover over a point. V/STOL (vertical/short takeoff and landing) aircraft such as the Harrier "Jump Jet" have been designed to combine the advantages of both types of vehicle. Helicopters with short wings to improve their efficiencies over long distances have also been tested, but so far neither these nor the V/STOL aircraft have fully met the needs of their potential users. The Bell Augusta BA-609, which incorporates tilting rotors at the end of each wing, is currently under development for commercial uses (fig. 145).

TECHNOLOGY AND INNOVATION

At a more fundamental level, many new technologies are being studied that could revolutionize our concept of flight. These could radically affect all aspects of aerospace design, including control systems, human/machine interfaces, and safety measures.

A bird changes the shape of its wings in order better to control its movement at various speeds and types of flight, from soaring to landing. The muscles and tendons in the bird's wings cause the wing shape to change, from almost straight out for slow speed to swept right back as the bird dives. Research is currently being undertaken that would allow the wings of airplanes to "morph," or change position, as those of a bird do. It is believed that this could increase the effectiveness

145. The Bell Agusta BA-609 civilian Tilt Rotor. This runway-independent
aircraft can land and hover like a helicopter but has the speed and
performance of a small airplane. The rotors at the end of each wing tilt
during flight to change the vehicle's mode from vertical to horizontal
(helicopter to airplane).

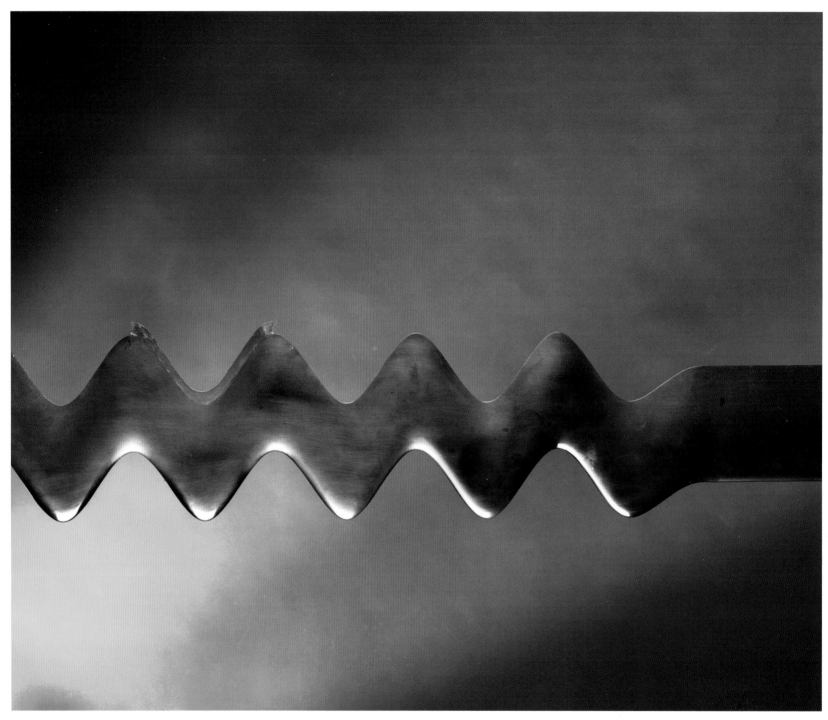

of wings at all speeds. Present-day airplane wings have slats, elevons, ailerons, flaps, and other surfaces, both to control the aircraft and to produce additional lift on take-off and landing. Slats are the metal pieces that drop forward on the leading edge of the wings, while flaps are the surfaces at the rear of the wing that rotate down at various angles during take-off and landing, causing the shape of the wing to change. These flaps currently require separately built actuators to move the surfaces. The morphing wing would have a bone-like structure

for support, covered by a flexible skin with embedded muscle-like actuators. The new wing would not just bend to act like present-day flaps but would also change thickness or airfoil shape along with the span and sweep. The result would be increased efficiency, greater speed, and reduced cost (fig. 148).

Work is being done, too, on self-healing materials and systems. Imagine materials that, when damaged, could reform themselves into their original shape. These "smart" materials would be

146. A model of a wavy-blade helicopter rotor, c. 1999. The familiar, loud slapping sound a helicopter makes as it flies overhead is a nuisance to civilians on the ground and those inside the helicopter; it is also a danger to military helicopters that are trying to operate undetected. In an effort to change or reduce this slapping sound, made by the blade as it rotates through the air, researchers have developed several unconventional-looking helicopter blades. This example, tested in the Transonic Dynamics Tunnel (TDT) in 1999 and still under development, is one that has resulted in reduced rotor noise.

147. A wind tunnel model of scalloped wing, c. 2001. This scalloped leading-edge wing got its inspiration from such marine creatures as the hammer-head shark and the humpback whale. Nature often provides improvements in efficiency that are not obvious to engineers. The hope is that this configuration will reveal another of nature's secrets. Modern engineers are trying to determine which benefits the sometimes unique configurations in nature could bring to man-made systems.

able to remold themselves physically into their original configuration. Related to these are systems that use neural networks, enabling an "intellegent" vehicle to learn from previous experience and to adapt its path if a part of the system fails. Software is currently being tested that can allow damaged aircraft to continue flying and land safely. This new software would use the remaining functional systems to bypass the failed components. Integral to both of these concepts is the "smart" sensor. Located within the vehicle, such sensors would monitor the vehicle's functions and transmit this data to a central processor, pilot, or ground control.

METHODS OF PROPULSION

All vehicles need a propulsion system to make them go. Even human-powered aircraft have a system: the person that propels the vehicle through the air. Many new methods of vehicle propulsion are being studied, ranging from high-speed air-breathing systems to improved jet engines and advanced general aviation engines (figs. 149, 150). The goal is to increase efficiency while lowering operational cost and greatly reducing noise and fuel consumption. These methods of propulsion should also be less environmentally damaging, producing smaller amounts of polluting emissions such as nitrous oxide and carbon dioxide. Current research

148. The end product of current "morphing" studies could be a futuristic vehicle like the Eagle Vision. Inspired by nature, it would be capable of changing shape to meet the needs of various flight conditions and speeds.

149. Artist's concept of the NASA
X-43A, an uncrewed test-bed for
scramjet engine technologies.
The X-43A is air launched from a
modified NB-52 and is boosted to
Mach 7 using a modified Pegasus
booster.

projects include a chevron-shaped nozzle mounted to engine exits to reduce noise (fig. 152). Other areas of investigation include the process of fuel combustion, and new materials that might improve engines' efficiency and environmental quality.

Carbon-based fuels—jet fuels or kerosene—are currently the means of propelling an aircraft, but studies also include the possibility of solar power. Since the world is not always lit, however, a way to store the sun's energy is required. The answer may be the fuel cell, which stores solar energy and converts it as required into electricity to power the motors. This concept is not new and has been tested for both automobiles and aircraft, but the problem is efficiency. New hydrogen–oxygen regenerative fuel cells are being

developed that would increase storage capacity. They are based on a proton-exchange membrane, which would electrolyze water into its basic gases of hydrogen and oxygen. These gases would be stored under pressure and later recombined in the fuel cell to produce electricity (fig. 151).

THE COCKPIT

The cockpit of the future will no doubt feature many of the advanced technologies that are being developed today, and others that can only be imagined. "Glass cockpits" are already becoming a reality, and have been integrated into the Boeing 777 (which entered service in 1995) and the Space Shuttle orbiter. A glass cockpit uses computer touchscreens to display information formerly shown by means of dials, gauges, and other analog systems. Advanced sensor suites

150. The Hyper X (X-43A) high-speed air-breathing scramjet engine being tested in the 8 Foot High-Temperature Tunnel at NASA Langley. The Hyper X uses cutting-edge scramjet technology to power the craft at altitudes and speeds previously impossible with air-breathing engines.

152. The "Chevron Nozzle" was designed to reduce jet noise from modern turbofan and turbojet engines. A series of nozzles has been ground- and flight-tested at NASA.

151. The Helios Prototype, an unmanned solar-powered flying wing, during a test flight over the Pacific Ocean. The Helios was designed to develop and test the technology necessary for long-duration high-altitude uncrewed flight. In 2001 Helios reached an altitude of 98,600 feet (30,053 meters), making it the highest uncrewed solar-powered flight in history.

154. (left) An advanced cockpit concept for commercial aircraft. The cockpit of the future currently under development combines advanced avionics and communications technologies with ergonomic displays, including glass cockpits and head-up displays.

153. (above) The cockpits of the Space Shuttle orbiters are in the process of being upgraded. These upgrades include a new "glass cockpit," in which many of the standard gauges and dials are replaced with modern computer displays with touch screens.

and computer uplink—like having the internet in the cockpit—would allow the transfer of up-to-the-minute weather and condition data to the pilot, thus increasing the safety of the aircraft (figs. 153, 154). New "synthetic vision" systems would allow a pilot to "see" in any weather, via displays overlaid with computer-generated terrain maps, with inputs from multiple sensors both on the aircraft and on the ground (figs. 155, 156).

Another new technology being studied for the control of aircraft is that of neural control. Instead of a control stick or yoke, this system uses a series of sensors mounted on a glove worn by the pilot. The pilot moves his or her hand as though holding a control stick, and the system interprets the neural commands made by the muscles' movement, issuing commands to the aircraft just as an actual control stick would. This system has been shown to be effective in simulator flights. In the long term, perhaps we might even see the neural control of aircraft by thought.

With so much technological innovation on the horizon the question arises, Will there be the need for human pilots in the future? This in turn raises the question, Will people trust their lives to a machine? Whatever the answer, great advancements in the development of unpiloted aerial vehicles have been made in recent years. The military is currently operating uncrewed vehicles for dangerous surveillance missions over enemy territory. NASA has looked at these vehicles for testing future concepts and for long duration missions. The uncrewed systems put people out of harm's way during the mission; they can also serve to reduce the size and complexity of the aircraft, due to the fact that it does not have to support a pilot during the flight. NASA has successfully tested the ability of a commercial grade airplane to be flown without a pilot (fig. 157).

FLIGHT ON OTHER WORLDS

The concept of flight—even when limited to that involving airplanes and balloons—is not confined to the earth's atmosphere. During the 1990s, with the success of the Mars Rover (which crawled over the planet's surface, using onboard instruments to take pictures), new methods that would allow a more detailed large-scale look at Mars were analyzed. The idea of either a balloon or an airplane flying in the Martian atmosphere was initially proposed by NASA scientists to take place in 2003, the hundredth anniversary of

155, 156. A prototype "synthetic vision" system is shown retrofitted to a NASA 757 test aircraft. The system combines head-up and computer-enhanced displays. With head-up displays, information for the pilot (such as speed, altitude, etc.) can be projected on to a sheet of transparent material in his line of sight as he looks through the windscreen.

powered flight. The Mars airplane was to be a secondary payload on an exploration mission. Unfortunately this was not to be. During the last few years the concept of flight on Mars has again been studied, but this time as a primary mission. The change from secondary to primary allows for a larger, more complex, and more capable craft (fig. 159).

How can an aircraft or balloon fly on Mars? Earlier missions have shown that from an aerodynamic standpoint the Martian atmosphere is similar to that of high altitudes here on earth. It follows that a vehicle that can fly at the earth's high altitudes would be suited for Mars. A proposed Mars

157. The X-45A unmanned combat aerial vehicle (UCAV) in flight over Edwards Air Force Base. Unmanned aerial vehicles such as this are becoming the aircraft of choice for experimental testing, and are used in place of crewed aircraft for dangerous or tedious missions.

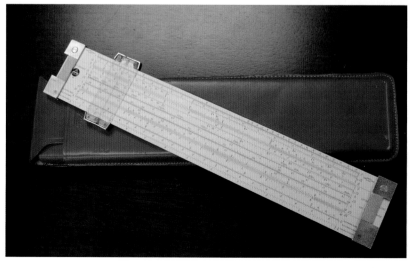

158. The slide rule—for decades the tool of the engineer and scientist—has today been replaced by the hand-held calculator and computer. In aerospace design, slide rules were used for basic calculations until the 1960s.

airplane was test-flown on earth in 2002 by launching it from a balloon at high altitude. The balloon itself—the oldest form of human flight—is the focus of other proposals. These high-tech balloons are nothing like their early predecessors, however, but modern marvels of science with multi-chambers and multi-gases (*e.g.* helium and hot air) that can perform a range of functions.

TOOLS OF THE TRADE

Until fairly recently, the methods employed to design aircraft (and, latterly, spacecraft) had changed little over the years. From the early twentieth century until the 1960s, pencil and paper were used to layout the vehicle while the age-old slide rule was employed for the basic engineering calculations (fig. 158). (Today few know what a slide rule is, let alone how to use one.) With the advent in the 1970s of smaller (not yet truly personal) computers, the engineer's methods and tools progressed to the next step.

At the same time the four-function calculator came into use, replacing the slide rule as the standard tool of the engineer. Today these four-function calculators, which once practically covered a desktop, have been replaced with multifunction, programmable calculators with more memory and power than most computers had a mere decade ago—and now they are the size of credit cards.

The advent of the personal computer has had the greatest impact of all upon the engineer's and designer's job. No longer is the designer bent over a large drafting table with enormous sheets of drafting paper, a handful of pencils or pens, and a mechanical drafting set consisting of various protractors, compasses, scales, rulers, and templates. These former tools of the design trade have been replaced by the computer mouse: nearly all design today is done using computer-aided design (CAD) software. These

159. Flight on other worlds is not just a dream: Mars, like earth, has an atmosphere in which an airplane could fly. NASA and industry are working together on concepts for a Mars airplane that could be a means to explore the planet's surface.

ARES

Aerial Regional-scale Environmental Survey of Mars

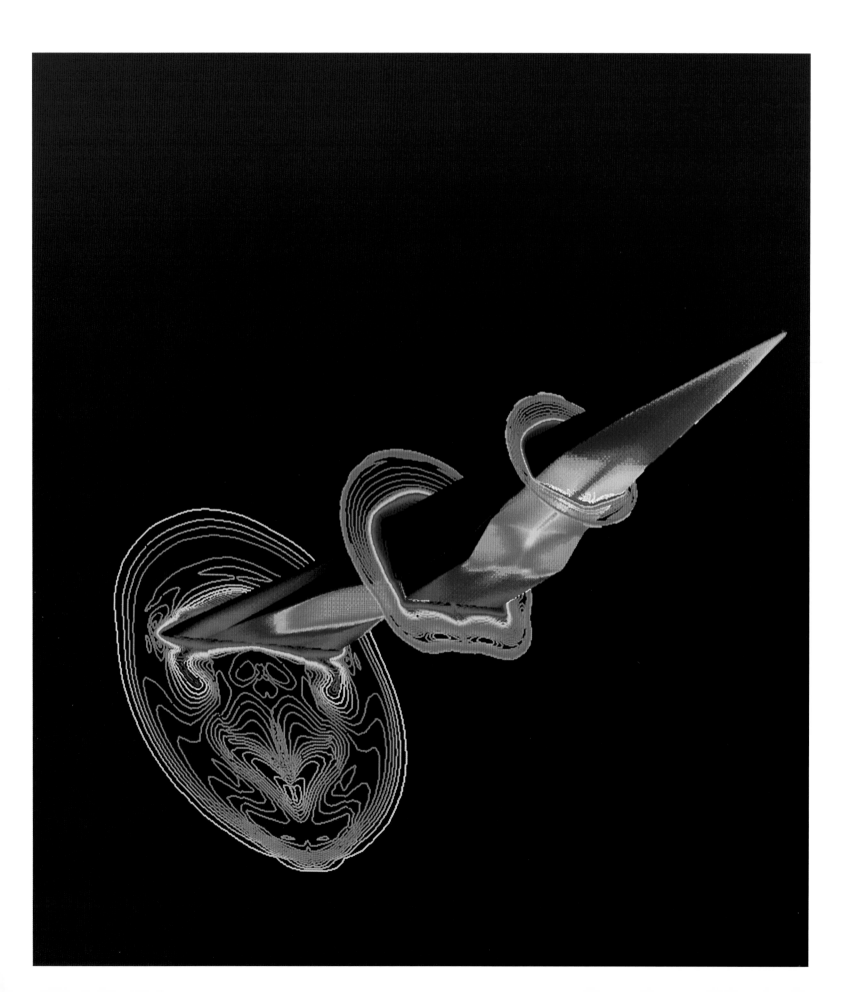

160. Computational fluid dynamic (CFD) analysis of a scramjet-powered aerospace vehicle at Mach 10. Using high-powered computers, CFD allows designers to visualize any type of fluid flow using the basic equations of fluid dynamics.

161. Computer simulation and analysis of the flowpath of fluids through a jet engine.

software programs are capable of automating routine tasks and of making changes in a few moments that would once have required the draftsperson to start from a clean sheet of paper. The Boeing 777 aircraft is the first aircraft to have been designed solely on computer, from the first drawings to the fit check of all the parts.

The computer is now an integral part of all aspects of the design process. The structure of a vehicle was once analyzed by hand, the vehicle was built and tested, and the structure subsequently modified and re-tested if a flaw was discovered. This process of structural analysis now takes place using computer software called NASTRAN, an updated version of a 1970s package jointly developed by NASA and industry. The software performs all the functions required to calculate the loads on a structure, replacing the often tedious hand calculations that were necessary in the past. Another job that has been changed by the computer is that of the aerodynamicist, who studies the flow of air over the vehicle or the flow of fluids through the engine. These can now be analyzed using computational fluid dynamics (CFD) software (figs. 160, 161).

One tool of the future that is in the works today is that of three-dimensional simulation, what many call "virtual reality." This has widespread applications, including flight training and the design of new systems and vehicles. One current 3-D simulation facility is Future Flight Central, which analyzes airport traffic. The software can be used to study different flight patterns or methods of increasing traffic while improving passenger safety. Such simulations help planners to understand and improve the flow of traffic in and out of airports, on the ground, and across the country. In the future these simulations may develop into the 3-D holograms seen in science fiction programs (fig. 162).

VISION OF THE FUTURE

A comparison of the visions of the future of aerospace put forth by the USA and Europe reveals that the aims of both are quite similar; the main difference is in terms of who will be the pre-eminent provider for the next century. Americans and Europeans alike have set themselves the goal of a system that is safer, more environmentally friendly, and less stressful for its users. This means a system that has higher capacity, shorter flight times, and fewer delays,

162. Future Flight Central is a "virtual" control tower allowing researchers and operators to test new procedures and processes that might ease the operations at airports. The facility can be modified to simulate any airport in the world.

with an increased level of comfort in the aircraft. The US plan, *The NASA Aeronautics Blueprint— Toward a Bold New Era of Aviation*,[6] includes discussions of digital airspace, revolutionary vehicles, security and safety, and the need for an educated workforce; while the European plan, *European Aeronautics: A Vision for 2020*,[7] looks at responses to society's needs, the question of global leadership, public policy and regulation, and how to identify research agenda. Both plans cover similar ground and have at their core the same vision for society.

As aircraft have become commonplace, the public's view of flight has changed from something approaching awe to seeing it as just of another mode of transportation. As we begin the second century of flight, we enter a new realm of possibility—the difference between aircraft and spacecraft is set to all but disappear. The current configurations of aircraft may be relegated to the history books, along with those of the early pioneers. New configurations will integrate the latest technologies, making them safer, more efficient, and more economical. The only limits to the possibilities of flight are those of the human imagination.

Notes

1. "The X-Vehicles: Advancing the Limits of Technology," presentation made by Dr. Richard P. Hallion, May 21, 2002, Santa Clara, California, at the NASA-USAF-AIAA X-Vehicles Symposium. Available at www.aiaa.org.

2. In 2001 there were 17,150,000 departures for scheduled commercial aircraft, logging 34,060,000 flight hours. During this time there were a total of 417 fatalities, of which 399 were on board the aircraft. These data do not include hostile actions. The accident rate for the last ten years is 1.18 per million departures. Source: *2001 Statistical Summary*, Seattle (Boeing Commercial Airplanes Group) June 2001.

3. *Future Flight: A Review of the Small Aircraft Transportation System Concept*, Washington, D.C. (National Academy Press) 2002.

4. The speed of sound—sonic speed, or Mach 1—is a function of pressure and temperature, and varies with altitude. The speed of sound is approximately 760 miles (1223 km) per hour at sea level and 660 miles (1062 km) per hour at an altitude of 35,000 feet (10,668 meters).

5. Concorde travels at a speed of a little over Mach 2, or 1322 miles (2127 km) per hour, and carries about one hundred passengers; the Boeing 747-400 cruises at Mach 0.85, or 609 miles (980 km) per hour, and carries about 475 passengers.

6. *The NASA Aeronautics Blueprint—Toward a Bold New Era of Aviation*, Washington, D.C. (Government Printing Office) 2002.

7. *European Aeronautics: A Vision for 2020*, Luxembourg (Office for Official Publications of the European Communities) 2001.

163. The end of an era: in its last resting place, the HL-10 lifting body aircraft stands guard at the gate of the NASA Dryden Flight Research Center. The HL-10 was one of five heavy-weight lifting body concepts flown between July 1966 and November 1975 to study the concept of using a vehicle's body to generate lift rather than its wings. The lifting bodies were designed to test concepts for future entry vehicles, such as Space Shuttle.

AUTHOR BIOGRAPHIES

JOHN ZUKOWSKY

Dr. John Zukowsky is the John H. Bryan Curator of Architecture at The Art Institute of Chicago. He earned his Master's and Doctoral degrees from the State University of New York at Binghamton in 1974 and 1977, respectively, where he studied architectural and art history. Formerly architectural archivist at the Hudson River Museum in Yonkers, New York, he has worked at The Art Institute of Chicago since 1978. While there he has been responsible for major exhibitions and books such as *Chicago Architecture: 1872–1922* (1987), *Chicago Architecture and Design: 1923–1993* (1993), *Karl Friedrich Schinkel: the Drama of Architecture* (1994), and *Japan 2000: Architecture and Design for the Japanese Public* (1997). He has also become one of the few specialists, in the United States if not the world, on the history of architecture and design for the aerospace industry, with his travelling exhibitions and books such as *Building for Air Travel: Architecture and Design for Commercial Aviation* (1996) and *2001: Building for Space Travel* (2001). He is currently a member of the Chicago Centennial of Flight Commission.

TOM D. CROUCH

Dr. Tom Crouch is Senior Curator of the Division of Aeronautics at the National Air and Space Museum (NASM), Smithsonian Institution, Washington, D.C. A Smithsonian employee since 1974, he has served both the NASM and the National Museum of American History (NMAH) in a variety of curatorial and administrative posts. He holds an MA from Miami University (1968) and a Ph.D. from the Ohio State University (1976), both in history. In addition, he holds the honorary degree of Doctor of Humane Letters, conferred in June 2001 by the Wright State University. Tom Crouch is the author or editor of a number of award-winning books, including *The Bishop's Boys: A Life of Wilbur and Orville Wright* (W.W. Norton, 1989) and *Eagle Aloft: Two Centuries of the Balloon in America* (Smithsonian Institution Press, 1983). In the fall of 2000, President Clinton appointed Dr. Crouch to the chairmanship of the First Flight Centennial Federal Advisory Board, an organization created to advise the Centennial of Flight Commission on activities to commemorate the 100th anniversary of powered flight.

DOMINICK A. PISANO

Dr. Dominick A. Pisano is Chair of the Aeronautics Division at the National Air and Space Museum, Smithsonian Institution, Washington, D.C. Formerly Curator of Art at the Smithsonian, he has served the museum for twenty-six years. During this time he has been involved in numerous research and writing projects, as well as in collections management, exhibitions, and service to the public. He received his Ph.D. in American Studies from George Washington University in 1988 and his MS from The Catholic University of America in 1974. Dr. Pisano is the author of *To Fill the Skies with Pilots: The Civilian Pilot Training Program, 1939–1946* (University of Illinois Press, 1993), and is co-author of *Legend, Memory and the Great War in the Air* (University of Washington Press, 1992) and *Charles Lindbergh and the Spirit of St. Louis* (Harry N. Abrams, Inc., 2002).

JOHN D. ANDERSON, JR.

Dr. John Anderson, Jr. is Curator for Aerodynamics at the National Air and Space Museum, Smithsonian Institution, Washington, D.C. He is a graduate of the University of Florida and Ohio State University, where he gained a Ph.D. in Aeronautical and Astronautical Engineering in 1966. During his career he has acted as Lieutenant and Task Scientist at the Aerospace Research Laboratory at Wright-Patterson Air Force Base and Chief of the Hypersonic Group of the US Naval Ordnance Laboratory. In 1973 he became Chairman of the Department of Aerospace Engineering at the University of Maryland. In 1999 he retired from the University of Maryland and was appointed Professor Emeritus. Dr. Anderson's books include the award-winning *A History of Aerodynamics and Its Impact on Flying Machines* (Cambridge University Press, 1997) and *Aircraft Performance and Design* (McGraw-Hill, 1999). He is a Fellow of the American Institute of Aeronautics and Astronautics (AIAA), the Royal Aeronautical Society, London, and the Washington Academy of Sciences. He was awarded the John Leland Atwood Award (1989), the Pendray Aerospace Literature Award (1995), and the 2000 von Karman Lectureship in Astronautics by the AIAA.

JAMES R. HANSEN

Dr. James R. Hansen is a Professor of History at Auburn University in Alabama. He has published books and articles on a wide variety of topics in the history of science and technology, ranging from the early days of aviation through the first nuclear fusion reactors, the moon landings, and the environmental history of golf courses. NASA nominated his last book, *Spaceflight Revolution* (1995), for a Pulitzer Prize. Hansen's six-volume documentary history of aerodynamics, *The Wind and Beyond*, will be published by NASA from 2003 onward as part of the Wright brothers' centennial celebration. His book *The Bird is on the Wing: Aerodynamics and the Progress of the American Airplane* is due to appear in 2003 as part of the Texas A&M University Press Centennial of Flight Series. Dr. Hansen is currently at work on the first authorized biography of Neil A. Armstrong.

MICHAEL H. GORN

Dr. Michael H. Gorn is Chief Historian of the NASA Dryden Flight Research Center, Edwards, California. He received his doctorate in history from the University of Southern California in 1978, and in succeeding years has worked in various capacities as a public historian. Gorn worked for the Department of Defense (US Air Force) for thirteen years, as a staff historian, command historian, and finally as the Deputy Air Force Historian. He also served as the first historian of the US Environmental Protection Agency in Washington, D.C. He has been employed by NASA since 1996. Gorn is the author of a number of books about aeronautics and space flight, most notably *The Universal Man: Theodore von Karman's Life in Aeronautics* (the Smithsonian Institution Press, 1992) and *Expanding the Envelope: Flight Research at NACA and NASA* (The University Press of Kentucky, 2001). He is the recipient of the Alfred V. Verville Fellowship in Aerospace History from the National Air and Space Museum and was selected for the American Historical Association Aerospace History Fellowship. He presently serves on the Education Committee of the US Centennial of Flight Commission.

DENNIS R. JENKINS

Dennis R. Jenkins is an engineer who has worked on projects such as Space Shuttle and X-33 for the past twenty years. He is currently a consultant to various major aerospace companies, primarily in the Cape Canaveral, Florida area. Jenkins has also written aerospace histories, such as *Space Shuttle: The History of the National Space Transportation System* (Specialty Press, 3rd edition 2001). Other works include *Hypersonics Before the Shuttle: A Concise History of the X-15 Research Airplane* (2000), and more than twenty-five other books on modern military aircraft.

ANTHONY M. SPRINGER

Anthony M. Springer is the director for all NASA Centennial of Flight Activities and the Alliance Development Manager for the Office of Aerospace Technology at NASA Headquarters. He is also a member of the US Centennial of Flight Commission History and Education Committee. He has served as the NASA resident manager for the X-34 project, and as a project and test engineer with NASA Marshall Space Flight Center. Anthony Springer is the co-editor of *Flight: A Celebration of 100 Years in Art and Literature* (2003) and has contributed to numerous other technical and historical publications. He is a graduate of the University of Illinois.

GLOSSARY

Most of the terms in this glossary are simplified from either the *Dictionary of Technical Terms for Aerospace Use*, ed. William H. Allen, Washington, D.C. (Government Printing Office) 1965, or from the *NASA Aeronautical Dictionary* by Frank Davis Adams, Washington, D.C. (Government Printing Office) 1959. The *Dictionary of Technical Terms* has been updated and placed on the web by Daniel R. Glover, Jr. of the NASA Glenn Research Center. It is available on the web at: http://roland.lerc.nasa.gov/~dglover/dictionary//content.html

ABLATING MATERIAL

A material designed to provide protection from heat by the removal or loss of (liquid or vapor) mass thus blocking the transfer of heat from the rest of the vehicle.

ACCELERATION

The rate of change of velocity.

ACCELEROMETER

A instrument that measures acceleration or gravitational forces capable of imparting acceleration.

AERODYNAMIC FORCE

The force exerted by a moving gaseous fluid upon a body completely immersed in it.

AERODYNAMICS

The science that deals with the motion of air or other gaseous fluids and the forces acting on bodies moving through such fluids.

AEROELASTICITY

Any phenomenon that includes the mutual interaction between aerodynamic loads and structural deformation.

AEROSPIKE ENGINE

A rocket engine in which the outside or free-stream air is used as the rocket nozzle. This engine is believed to be more efficient than current rocket engines since its exhaust gases will always be expanded optimally to the free-stream conditions.

AEROTHERMODYNAMICS

The study of aerodynamic phenomena at gas velocities sufficiently high that the thermodynamic properties of the gas become important.

AILERON

A movable control surface or device, one of a pair or set located in or attached to the wings on both sides of an airplane. The primary usefulness of ailerons is to control the airplane laterally or in roll by creating unequal or opposing lifting forces on opposite sides of the airplane. An aileron commonly consists of a flap-like surface at the rear of a wing.

AIR-BREATHING

When applied to an engine, a type that takes in air for the purpose of combustion.

AIRCRAFT

Any structure, machine, or contrivance, especially a vehicle, designed to be supported by the air, being borne up either by the dynamic action of the air upon the surfaces of the structure or object, or by its own buoyancy. It is most often applied to craft designed to support or convey a burden in or through the air.

AIRFOIL

A structure, piece, or body, originally likened to a foil or leaf in being wide and thin, designed to obtain a useful reaction on itself in its motion through the air.

AIRFRAME

The assembled structural and aerodynamic components of an aircraft or rocket vehicle that support the different systems and subsystems integral to the vehicle. The word *airframe*, carried over from aviation usage, remains appropriate for rocket vehicles since a major function of the airframe is performed during flight within the atmosphere.

ATMOSPHERE

The envelope of air surrounding the Earth; also a unit of pressure equal to 14.7 pounds per square inch (1.033 kg per square centimeter).

ATMOSPHERIC PRESSURE

The pressure at any point in an atmosphere due solely to the weight of the atmospheric gases above the point concerned.

BALLISTIC BODY

A body free to move, behave, and be modified by ambient conditions, substances, or forces, as by the pressure of gases in a gun, by rifling in a barrel, by gravity, by temperature, or by air particles. A rocket with a self-contained propulsion unit is not considered a ballistic body during the period of its guidance or propulsion.

BALLISTICS

The science that deals with the motion, behavior, and effects of projectiles, especially bullets, aerial bombs, rockets, or the like; the science or art of designing and hurling projectiles so as to achieve a desired performance.

BOUNDARY LAYER

The layer of fluid in the immediate vicinity of a bounding surface; in fluid mechanics, the layer affected by viscosity of the fluid, referring ambiguously to the laminar boundary layer, turbulent boundary layer, planetary boundary layer, or surface boundary layer. In aerodynamics the boundary-layer thickness is measured from the surface to an arbitrarily chosen point, *e.g.* where the velocity is 99 per cent of the stream velocity.

BIPLANE

An airplane having double-decked wings, one above the other. In most configurations either the lower wing, or both the lower and upper wings, are divided by the fuselage.

BUFFETING

The beating of an aerodynamic structure or surfaces by unsteady flow, gusts, etc.; the irregular shaking or oscillation of a vehicle component owing to turbulent air or separated flow. *Onset* refers to the point in flight at which this condition occurs.

CFD (COMPUTATIONAL FLUID DYNAMICS)

The use of a computer to process large numbers of computations, utilizing the basic equations of physics to calculate the flow of fluid either internal or external to a body.

CHINE

Tapers on the side of an aircraft, such as those on the SR-71.

CHORD

A straight line intersecting or touching an airfoil profile at two points. This line is usually a datum line joining the leading and trailing edges of an airfoil. The chord length is the length of the chord of an airfoil section between the extremities of the section.

COMPRESSOR

A machine for compressing air or other fluid. Compressors are distinguished (i) by the manner in which fluid is handled or compressed, *e.g.* the axial flow, centrifugal, double-entry, free-vortex, mixed-flow, single-entry, and supersonic compressors; or (ii) by the number of stages, as with the multistage or single-stage compressor.

DFBW (DIGITAL FLY-BY-WIRE)

An aircraft that uses a computer connected to digital cables to transmit commands to a vehicle's control surfaces, instead of a system of physical hydraulic tubing or cables to move those surfaces.

DRAG (SYMBOL D)

A retarding force acting upon a body in motion through a fluid, parallel to the direction of motion of the body. It is a component of the total fluid forces acting on the body. See *aerodynamics*.

FREE-FLIGHT FACILITY

A test facility in which small models are shot down a range or tube to determine the models' aerodynamic characteristics.

GRAVITY (SYMBOL G)

Viewed from a frame of reference fixed in the Earth, the force imparted by the Earth to a mass that is at rest relative to the Earth. Also a unit of acceleration, equal to the acceleration of gravity: 980.665 centimeter-second-squared, approximately 32.2 feet per second per second at sea level; used as a unit of stress-measurement for bodies undergoing acceleration.

GUIDANCE

The process of directing the movements of an aeronautical vehicle or space vehicle, with particular reference to the selection of a flight path.

GYRO

A device that utilizes the angular momentum of a spinning mass (rotor) to sense angular motion of its base about one or two axes orthogonal to the spin axis. Also called a gyroscope.

HYPERSONIC

Pertaining to speeds of Mach 5 or greater.

HYPERSONIC FLOW

In aerodynamics, the flow of a fluid over a body at speeds much greater than the speed of sound, and in which the shock waves start at a finite distance from the surface of the body.

INERTIAL GUIDANCE

Guidance by means of the measurement and integration of acceleration from within the craft.

JET ENGINE

An aircraft engine that derives all or most of its thrust from its ejection of combustion products (or heated air) in a jet, and that obtains oxygen from the atmosphere for the combustion of its fuel (or outside air for heating, as in the case of the nuclear jet engine); in this sense it is distinguished from a rocket engine. A jet engine of this kind may have a compressor (commonly turbine-driven) to take in and compress air (a turbojet); or it may be compressorless, taking in and compressing air by other means (such as a pulsejet or ramjet).

LEMS (LUNAR EXCURSION MODULE SIMULATOR)

A simulator designed to test astronauts on the operation of the Lunar Excursion Module, simulating both flight and the landing of the Module under the 1/6 gravity of the moon.

LIFT (SYMBOL L)

The component of the total aerodynamic force that acts on a body perpendicular to the undisturbed airflow relative to the body.

LIFTING BODY

An aircraft that uses the lift generated by its body for flight, as opposed to a general aircraft, which uses its wings to generate the lift needed for flight.

MACH (SYMBOL M)

(Named after the Austrian scientist Ernst Mach, 1838–1916.) A number expressing the ratio of the speed of a body or of a point on a body to the local speed of sound.

MANOMETER

An instrument for measuring the pressure of gases and vapors both above and below atmospheric pressure.

MONOCOQUE

A type of construction, such as a rocket body, in which all or most of the stresses are carried by the skin.

MONOPLANE

An aircraft with a single wing.

NACELLE

A streamlined enclosure to accommodate an aircraft engine; not part of the fuselage.

NACELLE PYLON

The structure, usually below a wing, used to attach the engine nacelle and the engine to the wing.

ORBITER

In a reusable vehicle, the orbiter is the vehicle that travels to and from space carrying the crew. In the space transportation system, or Space Shuttle, the orbiter is the white craft containing the crew and payload that goes into orbit and subsequently returns to Earth and is reused.

PITCH

An angular displacement about an axis parallel to the lateral axis of the vehicle. To an observer sitting in the vehicle and looking forward, this motion would be in the up and down direction.

PLANFORM

The shape or form of an object, such as an airfoil, as seen from above, as in a plan view.

RAMJET ENGINE

A type of jet engine with no mechanical compressor consisting of a specially shaped tube or duct open at both ends, the air necessary for combustion being shoved into the duct and compressed by the forward motion of the engine. In the engine the air passes through a diffuser and is mixed with fuel and burned, the exhaust gases issuing in a jet from the rear opening. The ramjet engine cannot operate under static conditions.

REACTION CONTROL SYSTEM (RCS)

Systems of normally cold gas jets used to control or move a vehicle at high altitudes or in space, where aerodynamic control surfaces would not function.

REYNOLDS NUMBER (SYMBOL R)

(Named after the English scientist Osborne Reynolds, 1842–1912.) A nondimensional parameter representing the ratio of the momentum forces to the viscous forces in fluid flow.

ROLL

The act of rolling; the rotational or oscillatory movement of an aircraft or similar body about a longitudinal axis through the body.

ROTOR AIRCRAFT

An aircraft that uses a rotor or rotors for lift and propulsive force. In general this would be a helicopter.

SCRAMJET

A type of ramjet engine that uses supersonic combustion to reach higher speeds than a ramjet, possible since the air does not need to slow to subsonic speeds for combustion to take place.

SKUNK WORKS

The nickname for the Lockheed Martin advanced development group, founded by Clarence C. "Kelly" Johnson. It has designed such vehicles as the P-80, F-104, U-2, SR-71, and the F-117 stealth fighter.

NASA CENTERS

STABLE PLATFORMS

A gyroscopic device designed to maintain a plane of reference in space, regardless of the movement of the vehicle carrying the stable platform.

STEALTH AIRCRAFT

An aircraft designed in such a way as to reduce its radar cross-section, thus allowing it to approach a target "stealthily."

STREAMLINING

The process of smoothing a vehicle or object to a more aerodynamically efficient configuration. In reference to non-vehicle applications, the design of an object to have a more aerodynamic appearance.

SUBORBITAL

Lacking obtainment of orbit.

SUBSONIC FLOW

The flow of a gas or fluid, such air over an airfoil, at speeds lower than the speed of sound.

SUPERCRITICAL WING

A wing in which the airfoil has been designed to delay the drag rise that accompanies transonic airflow, increasing the performance of the wing.

SUPERSONIC FLOW

The flow of a fluid over a body at speeds greater than the speed of sound, and in which the shock waves start at the surface of the body.

THERMODYNAMICS

The study of the flow of heat.

TILT-WING AIRCRAFT

An aircraft in which the engines and propellers tilt, allowing the vehicle to take off as a helicopter and fly as an airplane.

TRANSONIC

At or near the speed of sound (Mach 1), or within the range of speed in which flow patterns change from subsonic to supersonic (about Mach 0.8 to 1.2).

TRIMOTOR

An aircraft with three engines.

TURBOFAN

A turbojet engine in which additional propulsive thrust is gained by extending a portion of the compressor or turbine blades outside the inner engine case. The extended blades propel bypass airflows along the engine axis but between the inner and outer engine casing. This air is not combusted but does provide additional thrust, caused by the propulsive effect imparted to it by the extended compressor blading.

TURBOJET ENGINE

A jet engine incorporating a turbine-driven air compressor that takes in and compresses the air for the combustion of fuel (or for heating by a nuclear reactor), the gases of combustion (or the heated air) being used both to rotate the turbine and to create a thrust-producing jet.

TURBOPROP

An engine consisting of a turbojet engine used to drive a propeller.

V/STOL (VERTICAL/SHORT TAKE-OFF AND LANDING)

Aircraft capable of short or vertical take-offs.

WIND TUNNEL

A tube-like structure or passage, sometimes continuous, in which a high-speed movement of air or other gas is produced (*e.g.* by a fan). Objects such as engines, aircraft, airfoils, rockets, or models of these objects are placed inside the tunnel in order to investigate the airflow about them and the aerodynamic forces acting upon them.

WIND-TUNNEL BALANCE

A device or apparatus that measures the aerodynamic forces and moments acting upon a body tested in a wind tunnel.

WING SPAN

The span of the wing, measured in a straight-line distance between the tips or outermost extremities.

YAW

An angular displacement about the lateral axis parallel to the longitudinal axis. For an observer sitting in the vehicle and looking forward, this motion would be from side to side.

In this publication numerous references are made to the National Aeronautic and Space Administration (NASA) and its predecessor, the National Advisory Committee for Aeronautics (NACA). The NACA was founded on March 3, 1915, by the United States Congress in order to supervise and direct the scientific study of the problems of flight with a view to their practical solution. The NACA succeeded and grew over the next fifty years until the advent of the space race and the launch of Sputnik by the Soviet Union. NASA was established on October 1, 1958, as the civilian organization responsible for both aeronautics and astronautics with the NACA as its core. Today NASA has ten centers whose research encompasses many new technologies, advancing the state of the art in everything from high-speed computers and air and space flight to nano- and biotechnologies. In addition to its accomplishments in space flight, NASA supports extensive research in the fields of aeronautics, space science, earth science, and biological and physical sciences. The original NASA aeronautical centers are mentioned extensively in this volume. Additional information on NASA can be found on the web at http://www.nasa.gov.

Langley Research Center in Hampton, Virginia, was originally established in 1917 as NACA's first field laboratory. It is known for its research into airframe and atmospheric systems, along with advanced structures and materials.

John H. Glenn Research Center at Lewis Field in Cleveland, Ohio, was originally established in 1941 as the Aircraft Engine Research Laboratory and later renamed the Lewis Research Center. It is recognized for its extensive experience in aero-propulsion and turbomachinery.

Ames Research Center in Sunnyvale, California, was established in 1940 as the Ames Aeronautical Laboratory at Moffett Field, California. Ames specializes in research into aviation operations systems and information technology.

Hugh L. Dryden Flight Research Center started in 1946 at Muroc Army Airfield (now Edwards), California, as the Muroc Flight Test Unit, later changing its name to the High-Speed Flight Research Station. It is responsible for the testing of new research aircraft, aircraft technologies, and atmospheric flight operations.

selected bibliography

Richard Sanders Allen, *The Northrop Story, 1929–1939*, New York (Orion Books) 1990

American Airport Design, Washington, D.C. (American Institute of Architects Press) 1990; originally published by Leigh Portland Cement Co., Allentown, Pa., 1930

John D. Anderson, Jr., *A History of Aerodynamics and Its Impact on Flying Machines,* Cambridge, UK (Cambridge University Press) 1997

John D. Anderson, Jr., *Aircraft Performance and Design*, Boston (McGraw-Hill) 1999

Walter J. Boyne, *Art in Flight: The Sculpture of John Safer*, New York (Hudson Hills Press) 1991

Donald J. Bush, *The Streamlined Decade*, New York (George Braziller) 1975

Donald D. Baals and William R. Corliss, *Wind Tunnels of NASA*, Washington, D.C. (Government Printing Office) 1981

Sheldon Chaney and Martha Cheney, *Art and the Machine*, New York (Whittlesey House) 1936, reprinted New York (Acanthus Press Reprint Series) 1992

Edward W. Constant, II, *The Origins of the Turbojet Revolution*, Baltimore, Md. (The Johns Hopkins University Press) 1980

A. Scott Crossfield, *Always Another Dawn: The Story of a Rocket Test Pilot*, North Stratford, N.H. (Ayer Company Publishers) 1960

Tom D. Crouch, *A Dream of Wings: Americans and the Airplane, 1875–1905*, New York (W.W. Norton) 1981

Tom D. Crouch, *The Bishop's Boys: A Life of Wilbur and Orville Wright*, New York (W.W. Norton) 1989

G. Eiffel, *The Resistance of Air and Aviation: Experiments Conducted at the Champs-de-Mars Laboratory*, trans. Jerome C. Hunsaker, London (Constable & Co.) and Boston (Houghton Mifflin & Co.) 1913

Herbert Fenster, *The $5 Billion Dollar Misunderstanding*, Annapolis, Md. (Naval Institute Press) 2001

Eugene S. Ferguson, *Engineering and the Mind's Eye*, Cambridge, Mass. (MIT Press) 1993

David Gartman, *Auto Opium: A Social History of American Automotive Design*, London (Routledge) 1994

Charles H. Gibbs-Smith, *Sir George Cayley's Aeronautics, 1796–1855*, London (HMSO) 1962

Charles H. Gibbs-Smith, *Aviation: An Historical Survey From Its Origins to the End of World War II*, London (HMSO) 1970

Michael H. Gorn, *Expanding the Envelope: Flight Research at NACA and NASA*, Lexington, Ky. (University Press of Kentucky) 2001

George W. Gray, *Frontiers of Flight: The Story of NACA Research*, New York (Alfred A. Knopf) 1948

Richard P. Hallion and Michael H. Gorn, *On the Frontier: Flight Research at Dryden*, Washington, D.C. (Smithsonian Institution Press) 2003

Paul A. Hanle, *Bringing Aerodynamics to America*, Cambridge, Mass. (MIT Press) 1982

James R. Hansen, *Engineer in Charge: A History of the Langley Aeronautical Laboratory*, Washington, D.C. (Government Printing Office) 1987

Peter Jakab, *Visions of a Flying Machine: The Wright Brothers and the Process of Invention*, Washington, D.C. (Smithsonian Institution Press) 1990

Dennis R. Jenkins, *Lockheed SR-71/YF-12 Blackbirds*, Volume 10 in the *WarbirdTech* Series, North Branch, Minn. (Specialty Press) 1997

Dennis R. Jenkins, *Lockheed Secret Projects: Inside the Skunk Works*, St. Paul, Minn. (MBI Publishing Company) 2001

Dennis R. Jenkins, *Space Shuttle: The History of the National Space Transportation System—The First 100 Missions*, Cape Canaveral, Fla. (Specialty Press) 2001

Dennis R. Jenkins and Tony Landis, *Hypersonic: The Story of the North American X-15*, North Branch, Minn. (Specialty Press) 2003

Raymond Loewy, *Industrial Design*, Woodstock, N.Y. (The Overlook Press) 1988; reprinted 2000

Laurence K. Loftin Jr., *Quest for Performance: The Evolution of Modern Aircraft*, Washington, D.C. (Government Printing Office) 1985

Pamela E. Mack (ed.), *From Engineering Science to Big Science: The NACA and NASA Collier Trophy Research Project Winners*, Washington, D.C. (Government Printing Office) 1998

Jeffrey L. Meikle, *Twentieth Century Limited: Industrial Design in America, 1925–1939*, 2nd edn, Philadelphia, Pa. (Philadelphia Temple University Press) 2001

Jay Miller, *Lockheed Skunk Works: The First 50 Years*, Hinckley, UK (Midland Publishing) 1995

Jay Miller, *The X-Planes: X-1 to X-45*, North Branch, Minn. (Specialty Press) 2001

Ronald Miller and David Sawers, *The Technical Development of Modern Aviation*, London (Routledge & Kegan Paul) 1968

Machine Art, exhib. cat., New York, The Museum of Modern Art, 1934; 60th Anniversary Edition, The Museum of Modern Art, 1994

Leslie E. Neville, *Aircraft Designers' Data Book*, New York (McGraw-Hill) 1950

Noel Pemberton-Billing, *The Aeroplane of Tomorrow*, London (Robert Hale Ltd.) 1941

Alan Pope and William H. Rae, *Low-Speed Wind Tunnel Testing*, 2nd edn, New York (John Wiley & Sons) 1984

K.G. Pontus Hulte'n, *The Machine as Seen at the End of the Mechanical Age*, New York (The Museum of Modern Art) 1968

N.H. Randers-Pherson, "Pioneer Wind Tunnels," in Smithsonian Miscellaneous Collections 93, Washington, D.C. (Government Printing Office) January 19, 1935

Daniel P. Raymer, *Aircraft Design: A Conceptual Approach*, Restin, Va. (American Institute of Aeronautics and Astronautics) 1989

R. Dale Reed with Darlene Lister, *Wingless Flight: The Lifting Body Story*, Washington, D.C. (Government Printing Office) 1997; reprinted by the University Press of Kentucky, 2002

Alex Roland, *Model Research: The National Advisory Committee for Aeronautics*, Washington, D.C. (Government Printing Office) 1985

Howard L. Scamehorn, *Balloons to Jets. A Century of Aeronautics in Illinois, 1855–1955*, Chicago (Henry Regnery Co.) 1957

James Hay Stevens, *The Shape of the Aeroplane*, London (Hutchinson & Co. Ltd) 1953

Loyd S. Swenson Jr., James M. Grimwood, and Charles C. Alexander, *This New Ocean: A History of Project Mercury*, Washington, D.C. (Government Printing Office) 1966

Henk Tenneckes, *The Simple Science of Flight: From Insects to Jumbo Jets*, Cambridge, Mass. (MIT Press) 1996

James E. Tomayko, *Computers Take Flight: A History of NASA's Pioneering Digital Fly-By-Wire Project*, Washington, D.C. (Government Printing Office) 2000

Milton O. Thompson, *At The Edge of Space: The X-15 Flight Program*, Washington, D.C. (Smithsonian Institution Press) 1992

Milton O. Thompson and Curtis Peebles, *Flying without Wings: NASA Lifting Bodies and the Birth of the Space Shuttle*, Washington, D.C. (Smithsonian Institution Press) 1999

Walter Vincenti, "The Retractable Airplane Landing Gear and the Northrop 'Anomaly': Variation-Selection and the Shaping of Technology," in *Technology and Culture*, vol. 35, 1994, pp. 1–33

Walter G. Vincenti, *What Engineers Know and How They Know It: Analytical Studies from Aeronautical History*, Baltimore, Md. (The Johns Hopkins University Press) 1990

The Machine Age in America 1918–1941, exhib. cat. by Richard Guy Wilson, Dianne H. Pilgrim, and Dickran Tashjiam, New York, Brooklyn Museum of Art, 1986; reprinted 2001

John Walter Wood, *Airports: Some Elements of Design and Future Development*, New York (Coward-McCann Inc.) 1940

Fred E. Weick and James R. Hansen, *From the Ground Up: The Autobiography of an Aeronautical Engineer*, Washington, D.C. (Smithsonian Institution Press) 1988

John Zukowsky (ed.), *Building for Air Travel: Architecture and Design for Commercial Aviation*, Munich (Prestel) 1996

John Zukowsky (ed.), *2001: Building for Space Travel*, New York (Harry N. Abrams) 2001

John Zukowsky, *Space Architecture: The Work of John Frassanito & Associates for NASA*, Stuttgart (Edition Menges) 1999

PICTURE CREDITS

All photographs reproduced are from the archives of NASA, with the exception of the following:

Air Force Flight Test Center History Office: 21, 22, 24, 86, 91, 98, 109, 113, 115; American Airlines: 34; John Anderson: 38; The Art Institute of Chicago: 11–15; Auburn University, the Rickenbacker Papers: 5; Bell Agusta: 145; Boeing: 7, 9; Chicago Historical Society: 4; Cooper-Hewitt, National Design Museum, Smithsonian Institution, New York: 29; Corbis: 30, 31; Eclipse Aviation: 141; Hagley Museum and Library Wilmington, Del.: 25, 28, 32; Kelly Johnson: 48; Kunstmuseum, Basle: 37; Tony Landis 110; J. Lawrence Lee: 53; Lockheed Martin: 16, 80; Colonel Robert R. McCormick Research Center: 6; Naval Historical Center, National Archives, Washington, D.C.: 8; National Air and Space Museum, Smithsonian Institution, Washington, D.C.: 17, 26, 39; Tony Springer: 158; Steinkamp/Ballogg Photography: 10; United Airlines: 40; Alvin S. White: 112; Wright State University: 54

Additional material was also produced for NASA by the following:

John Frassanito & Associates: 132–34, 142–44, 148; Peggy Hopkins: 27, 33, 35, 42, 49, 106, 107, 121, 122, 128, 136, 146, 147; Steve Lighthill: 149; Garry Qualls: 159; Tom Trower: 137; Tom Tschida: 47

INDEX

ACKNOWLEDGEMENTS

Aerospace Design, like any book, is not the work of a single individual but a group, each person contributing his or her own part to the project. This is never more true than in an edited volume such as this, and the Editor would like to thank the many people who have brought it to fruition. Each of the chapter authors has contributed their time and talents to explain the workings of the aerospace world and the processes of design. I would also like to thank Tom Dixon, who has been an instrumental part of this project since its inception and has served as the director of planning and acquisition. He has worked diligently to acquire the objects and photographs presented in the exhibit. The men and women of the office of Aerospace Technology, especially Jenny Kishiyama and Bob Pearce, have given much support and guidance throughout. John Zukowsky of The Art Institute of Chicago has contributed immensely by co-curating *Aerospace Design*, and I would also like thank Roger Launius of the National Air and Space Museum, formerly Chief Historian of NASA, for his inputs into the exhibit, and Burt Ulrich, Curator of the NASA Art Program, for his assistance in this endeavor. The History Office at NASA headquarters has been an invaluable resource throughout. I am very grateful to my wife, Emily, for her patience and for taking the time to review this work, and to Julian Honer and the others at Merrell who made this book a reality.

I would also like to thank the many people at the NASA centers, civil servants and contractors who have helped to make *Aerospace Design* a possiblity:

NASA HEADQUARTERS

Louise Alstork, Nadine Andreassen, Jennifer Davis, Collin Fries, Steve Garber, John Hargenrader, Terry Hertz, Phil Milstead, Jane Odom, and Glenn Smith

DRYDEN FLIGHT RESEARCH CENTER

Jennifer Baer-Riedhart, Andy Blua, Joseph Ciganek, Gary Cosentino, Jennifer L. Hansen, Gerald Keever, Tony Landis, Steve Lighthill, Pete Merlin, Tony Moore, Greg Poteat, Mike Relja, Frank Romo, Wane Shively, Tom Tschida, and Benjamin Villanueva

AMES RESEARCH CENTER

Roger Ashbaugh, Daniel Bencze, Brett Casadante, Jeffrey Cross, Jonas Dino, Dr. Gregory A. Dorais, Gaye Graves, Joseph Lamica, Ron Lamica, Daniel Petroff, Dr. James Ross, Ken Stocking, and Mark Sumich

MARSHALL SPACE FLIGHT CENTER

James Aaron, Henry Brewster, Dewey Brown, Adeline Byford, John Dumoulin, Alonzo Frost, Karen Sodomick, Brenda Torres-Hill, and Holly Walker

GLENN RESEARCH CENTER

James E. Braatz, David M. DeFelice, James H. Dittmar, Larry H. Gordon, Michael W. Henry, Dennis L. Huff, David J. Lowenfeld, Richard A. Manco, Donald T. Palac, Marvin G. Smith, Dr. Charles J. Trefny, and Richard P. Woodward

LANGLEY RESEARCH CENTER

Dr. Balakrishna, Charles Bobbitt, Dr. Colin Britches, Thomas Brooks, Mike Chambers, Stan Cole, David Coleman, Dr. Erik Conway, Juan Cruz, Donald Day, Jim Hallissy, Fred Howell, Pete Jacobs, Allen Kilgore, Donna Lawson, Mike Marcolini, John Meador, John Micol, Roy Neff, David Reubush, Greg Shanks, David Shaw, Dr. Richard Wahls, Bob Webster, Matt Wilber, William Yeager, and John Zalarick

First published 2003 by
Merrell Publishers Limited

Head office:
42 Southwark Street
London SE1 1UN

New York office:
49 West 24th Street
New York, NY 10010

www.merrellpublishers.com

in association with

National Aeronautics and Space Administration

Design and layout © 2003 Merrell Publishers Limited

A catalogue record for this book is available from the
Library of Congress

British Library Cataloguing-in-Publication data:
     Aerospace design : aircraft, spacecraft, and the art
     of modern flight
     1.Aerospace engineering 2.Airplanes – Design and
     construction 3.Space vehicles – Design and
     construction 4.Aeronautics
     I.Springer, Anthony M. II.United States. National
     Aeronautics and Space Administration
     629.1'3

ISBN 1 85894 207 1

Edited by Anthony M. Springer

Produced by Merrell Publishers
Copy-edited by Christine Davis
Index by Diana LeCore
Art direction by Bark, layouts by Kate Ward

Printed and bound in Italy

Jacket, front:
Research pilot William Dana takes a moment to watch
NASA's NB-52B cruise overhead after a research flight
in the HL-10 at NASA's Dryden Flight Research Center,
Edwards, California. "NASA and its predecessor, the
National Advisory Committee for Aeronautics (NACA),
have been pivotal contributors to aerospace design
since NACA's founding in 1915. Teamed with their
industry counterparts, NASA engineers conceived the
aesthetically pleasing Apollo Command Module, as well
as its ugly-duckling companion, the Lunar Module. There
has probably never been a more graceful spacecraft than
the Space Shuttle, or one more toy-like than the HL-10
lifting body. In pursuing the conquest of air and space,
the NACA and NASA have influenced the design of
aircraft and spacecraft the world over."

William H. Dana, April 1, 2003

Jacket, back:
A model of a proposed supersonic transport mounted in
the NASA Langley Full-Scale Wind Tunnel, 1975.

MERRELL
LONDON · NEW YORK